Mechanics, Boundary Layers, and Function Spaces

Diarmuid Ó Mathúna

Mechanics, Boundary Layers, and Function Spaces

Birkhäuser
Boston • Basel • Berlin

Diarmuid Ó Mathúna
Dublin Institute for Advanced Studies
Dublin, Ireland

Library of Congress Cataloging-in-Publication Data
Ó Mathúna, Diarmuid.
Mechanics, boundary layers, and function spaces / Diarmuid Ó Mathúna.
p. cm.
Includes bibliographical references.
ISBN 0-8176-3464-9 (alk. paper)
1. Elastic plates and shells. I. Title.
QA935.02 1989
624.1'776—dc20 89-36924

Printed on acid-free paper.

ISBN 0-8176-3464-9
ISBN 3-7643-3464-9

Camera-ready text provided by the author using the Compugraphics 3400.
Printed and bound by Edwards Brothers, Inc., Ann Arbor, Michigan.
Printed in the U.S.A.

9 8 7 6 5 4 3 2 1

Preface

Concern with the class of problems investigated in this monograph began for me as a graduate student at MIT (1958-62) when serving as research assistant to Professor Eric Reissner who initiated me into the subject and whose influence — whether directly or dialectically — is probably discernable in the contours of the work. My first attempt at a systematic derivation of the equations of shell theory was made while on a summer assistantship with Professor Norman Levinson in 1960. Beyond gaining a sobering realization of the complexities involved, I made little progress at that time.

In 1962-64 while a Temporary Member at the Courant Institute of Mathematical Sciences (NYU) I made a fresh start, while benefiting from my association and discussions with Professor Fritz John. With the conviction that the full integration of the equations with respect to the thickness coordinate, by means of the Legendre representations, must lead to a clarification of the position of the two-dimensional theory in its three-dimensional context, the necessary computations were completed during that period. Several years passed while I became reconciled with the thought that the material needed to be organized as a monograph. This was done during 1969-70 while at the NASA Electronics Research Center in Cambridge, MA.

The favorable reviews by both Professors Louis Howard and William Shack and the interest shown by Professor John Lewis led to the first issue of the work as a Communication of the Dublin Institute for Advanced Studies, Series A, No. 28, 1984.

Special acknowledgement is due to Ms. Andrea Barth-Goldstein who typed the entire script. The typesetting was performed at Arts & Letters, Inc., Brookline, MA., and fitting such unmanageable material into an attractive format is due to the combined gifts of Ann Kostant, Debra Rattet and Steven Sagas.

The present (1989) edition has facilitated some updating of the bibliography and the correction of some errors where again the cooperation of Ann Kostant has been graciously forthcoming.

Diarmuid Ó Mathúna

Institúid Árd-Léinn Bhaile Átha Cliath
Áth Cliath, 4.

Dublin Institute for Advanced Studies
Dublin, 4.

Bealtaine 1989

Summary - Abstract

In spite of the long history of the theory of elastic plates and shells, many features of the contraction from a three- to a two-dimensional theory remain obscure. The lingering ambivalence, as to what physical interpretation should be given to the displacement functions proper to a two-dimensional description, is due to the fact that it has not been established in what sense, if any, the elements determined by the contracted formulation approximate the corresponding quantities of the three-dimensional theory. The present work is aimed at clarifying for the static case, within the framework of the linearized theory of elasticity, the precise significance of the simpler theory, by means of an equally precise formulation of the complementary problem. This will be effected through a complete and systematic integration, with respect to the thickness coordinate, of the full three-dimensional equations.

The recursion relations associated with the Legendre polynomials facilitates the integration with respect to the thickness coordinate of the system of three-dimensional equations. The repeated application of these formulae leads ultimately to Legendre representations for all field quantities, while the recognition of the orthogonal character of these series in the integrated form of the equations, and also in the edge conditions, yields an equivalent two-dimensional description of the full three-dimensional problem. This alternate formulation reveals how the simpler theory appears in the dominant terms, and a closer inspection shows that the full problem resolves into two uncoupled problems—one coinciding with the simpler theory involving only the dominant effects, the other consisting of the complementary problem, describing the residual effects, also formulated in two-dimensional terms.

The integration of the field equations for the elastic strip is effected in Chapter 1. In the case of beams of constant thickness, the formulation of the complementary problem yields an infinite set of ordinary differential equations with *constant* coefficients with an appropriate set of two-point boundary conditions. The clarification of the underlying function space structure in this simplest case, provides a model for the more complex problems formulated later.

In the analysis of plate theory treated in Chapter 2, we suppress most secondary features. On effecting the uncoupling of the integrated formulation, we find that the principal effects of bending and stretching are exactly described by slightly modified forms of the generalized theory of plane stress and Reissner's theory of plate bending, respectively. For the corresponding complementary problems, the formulation is followed by the reduction of the governing equations to a relatively compact form.

In the treatment of shell theory in Chapter 3, where we are forced to adopt an approximate method of integration, we do not aim for a comparably complete analysis of the respective problems. Beyond formulating the boundary value problem for the residual effects, we do not make any further reduction. In the case of the system of equations governing the principal effects, we perform a partial reduction. However, our main aim in this case is the derivation of a first-order system of asymptotically valid constitutive relations, giving an adequate description of the dominant effects in the interior. This is achieved in the final section and completes our analysis.

Contents

To
Jürgen Moser

"Mathematical physics and pure analysis are not merely adjacent powers, maintaining good neighborly relations, they mutually interpenetrate and their spirit is the same."

Henri Poincaré

General Introduction

The theory of elastic plates and shells deals with a class of problems in Solid Mechanics characterized by the restriction that the structures concerned are thin, signifying that, in the typical geometric representation, the particular diameter measuring the thickness is substantially smaller than the other length scales of the configuration. There is a direct correlation between this common mensural feature and certain mechanical properties shared by such structures, namely an enhanced flexibility in the thickness direction and the occurrence of boundary layer phenomena reflecting the fact that the influence of some of the conditions applied to the edge surface is confined to a relatively narrow neighborhood of their area of application.

Although these problems are three-dimensional in nature, it has long been recognized that, in the majority of cases of practical interest, the main patterns of stress and deformation are adequately described by a comparatively simple two-dimensional system of equations with an appropriately contracted form of the boundary conditions. In this two-dimensional formulation there is no recognition, either explicit or implicit, of the thickness coordinate. This contraction is intimately connected with the Principle of Saint Venant [14], implications of which have been proved for certain classes of problems by Toupin [16] and Knowles [7].

In spite of the long history of the theory, many features of this contraction from a three- to a two-dimensional theory remain obscure. The lingering ambivalence, as to what physical interpretation should be given to the displacement functions proper to a two-dimensional description, is due to the fact that it has not been established in what sense, if any, the elements determined by the contracted formulation approximate the corresponding quantities of the three-dimensional theory. These and other related issues can be resolved only in a context that includes a precise formulation of the complementary problem describing the effects suppressed in the simpler theory. The present work is aimed at clarifying for the static case, within the framework of the linearized theory of elasticity, the precise significance of the simpler theory, by means of an equally precise formulation of the complementary problem. This will be effected through a complete and systematic integration, with respect to the thickness coordinate, of the full three-dimensional equations.

Part of the original motivation for the contracted formulation arose from the fact that, in many cases of practical interest, the specification of the stress distribution of the edge-segment of the bounding surface is limited to a knowledge of the stress-resultants and stress-couples, which, in terms of the distribution through the thickness, correspond to the zero-th and first moments taken with respect to the thickness coordinate. Guided by this indication in selecting a basis for exhibiting the dependence on the thickness coordinate, we consider only those representations in which the leading pair of coefficients coincide with the zero-th and first moments. Adding the requirement that the basis be orthogonal leads to the adoption of a resolution in terms of Legendre polynomials. It is then consistent with physical intuition to expect that the pursuit of such an expansion procedure lead to a formulation, from the leading terms of which the simpler theory should emerge.

The recursion relations associated with the Legendre polynomials facilitates the integration with respect to the thickness coordinate of the system of three-dimensional equations. The repeated application of these formulae leads ultimately to Legendre representations for all field quantities, while the recognition of the orthogonal character of these series in the integrated form of the equations, and also in the edge conditions, yield an equivalent two-dimensional description of the full three-dimensional problem. This alternate formulation reveals how the simpler theory appears in the dominant terms, and a closer inspection shows that by making the proper distinction between principal and residual effects, the full problem resolves into two uncoupled problems – one coinciding with the simpler theory involving only the dominant effects, the other consisting of the complementary problem, describing the residual effects, also formulated in two-dimensional terms. It is also evident that for the analysis of the latter problem the natural context is an infinite-dimensional linear function space. The demonstration of the boundary layer nature of the influence of the edge conditions in the solution of the complementary problem would then be equivalent to the validation of Saint Venant's Principle for the configuration concerned.

Rather than immediately launch into the full ramifications of the theory of plates and shells, we start by considering the model problem of the elastic strip which, while admittedly artificial, nevertheless plays the role of the "hydrogen atom" in the theory. The question then is that of relating the one-dimensional equations of classical beam theory to the two-dimensional field equations of a planar elastic medium. Although beam theory should strictly be related to a three-dimensional medium – since it is concerned with a specialized form of the flat plate – it is more instructive to begin with a detailed examination of the less ambitious problem and thereby observe how the transformation of the two-dimensional field equations into an equivalent one-dimensional formulation yields the simplest manifestation of the phenomena we wish to analyze. Certain aspects of the finer points of beam theory may perhaps be more realistically discussed within the semi-inverse procedure proposed by Saint-Venant [14], and subsequently treated in the book by Clebsch [3], and also in the lectures of Poincaré [12]: however, this method, which has been further developed in the work of Pearson [11], Voigt [18], Michel [10], Almansi [1], Love [9], Timoshenko [15] and Goodier [4], does not apply itself to the questions investigated here.

The integration of the field equations for the elastic strip is effected in Chapter 1 and the relative simplicity permits, without obscuring the main goal, the simultaneous investigation of certain secondary features, the most tedious of which are the effects due to thickness variation. In the case of beams of constant thickness, the formulation of the complementary problem yields an infinite set of ordinary differential equations with *constant* coefficients with an appropriate set of two-point boundary conditions. The clarification of the underlying function space structure in this simplest case, provides a model for the more complex problems formulated later.

In the analysis of plate theory treated in Chapter 2, we suppress most secondary features. On effecting the uncoupling of the integrated formulation, we find that the principal effects of bending and stretching are exactly described by slightly modified forms of the generalized theory of plane stress and Reissner's theory of plate bending, respectively. For the corresponding complementary problems, the formulation is followed by the reduction of the governing equations to their relatively compact final form.

In the treatment of shell theory in Chapter 3, where we are forced to adopt an approximate method of integration, we do not aim for a comparably complete analysis of the respective problems. Beyond formulating the boundary value problem for the residual effects, we do not make any further reduction. In the case of the system of equations governing the principal effects, we perform a partial reduction. However, our main aim in this case is the derivation of the first-order system of asymptotically valid constitutive relations, giving an adequate description of the dominant effects in the interior corresponding to the Kirchhoff moment-curvature relations of plate bending. This is achieved in the final section and completes our analysis.

The use of Legendre representations has appeared in the work of a number of authors dealing with various aspects of these problems, e.g. Horvay [5] in dealing with the elastic strip, Cicala [2], Poniatovskii [13] and Krenk [8], in their respective treatments of plate theory, and the analysis of shell theory by Hu [6] and Vekua [17]. However, the procedure followed here, which does not appear to have been exploited heretofore, makes evident the advantages of the Legendre polynomial representations in a manner that shows them particularly tailored for the problems considered.

References

1. Almansi, E., Roma, Accademi Lincei Rendicotti, ser. 6, t. 10, 1901.
2. Cicala, P., Giornio Genio Civile, vol. 97, Nos. 4, 6, 9, 1959.
3. Clebsch, A., *Theorie der Elasticitat fester Korper,* Leipzig, 1862.
4. Goodier, J.N., Transactions of the Royal Society of Canada, vol. 33, ser. 3, 1938.
5. Horvay, G., Quarterly of Applied Mathematics, vol. 15, pp. 65-81, 1957.
6. Hu, W. C. L., Technical Report No. 5, NASA Contract NAS 94 (06), Sw. R.I., San Antonio, Texas, 1965.
7. Knowles, J. K., Archive for Rational Mechanics and Analysis, 21, pp. 1-22, 1965.
8. Krenk, S., Journal of Applied Mechanics, Vol. 48, 1981.
9. Love, A.E.H., *A Treatise on the Mathematical Theory of Elasticity,* 4th Ed. C.U.P., Cambridge, 1927.
10. Michel, J.H., Quarterly Journal of Mathemtics, vol. 32, 1901.

11. Pearson, K., Quarterly Journal of Mathematics, vol. 24, 1889.
12. Poincaré, H., *Leçons sur la théorie de l'élasticité,* Paris, 1892.
13. Poniatovskii, V.V., PMM Applied Mathematics and Mechanics, vol. 26, No. 2, 1962.
14. de Saint Venant, B., (a) Mémoires des Savants Etrangers, t. 14, 1855.
(b) Journal de Mathématique (Liouville), ser. 2, t. 2, 1856.
15. Timoshenko, S., *Theory of Elasticity,* McGraw-Hill Book Co., New York, 1934.
16. Toupin, R.A., Archive for Rational Mechanics and Analysis, 18, 1965.
17. Vekua, I.N., *Theory of Thin Shells of Varying Thickness* (in Russian), Tbilisi, 1965.
18. Voigt, W., Theoretische Studien über die Elasticitatverhaltnisse der Krystalle: Gottengen Abhandlunger, Bd. 34, 1887.

"The language of truth is simple."
Euripides

Chapter One

Beam Theory and the Residual Effects in the Elastic Strip

Introduction

In order to treat the problem of the straight elastic beam within the framework of a two-dimensional formulation, we consider the boundary value problem for the plane figure defined by the side-view projection of the three-dimensional body. Corresponding to the midplane of the beam the figure has an axis of symmetry, called the centerline or axis, which immediately defines a pair of reference directions in the plane; namely, the axial and transverse directions respectively parallel and normal to the centerline.

Bounding the figure is a simple closed curve which falls naturally into four segments – a pair of edges alternating with a pair of faces. The edges are transverse rectilinear segments, one at each end of the axis: thus the centerline bisects the edges perpendicularly and does not meet the boundary at any other point. The complementary segments, each of which joins a pair of edge endpoints on the same side of the symmetry axis, are called the faces of the beam; these are mirror reflections of each other in the centerline. The metrical characteristics of the beam are the length measured parallel to the axis and the thickness measuring the transverse intercept which may vary along the axial length.

The figure which has a prescribed mass per unit area is composed of a material whose mechanical behavior is in accordance with the Cauchy-Green law for a two dimensional isotropic elastic medium. In addition to the constraining action of the boundary conditions applied along the respective segments of the bounding curve the material is also subject to an interior body force distribution induced by the force field in which it lies. The only restriction, namely that the applied forces be consistent with overall static equilibrium, will follow as a consequence of the microscopic equilibrium equations and need not be considered as an extraneous condition.

The boundary conditions may take various forms which can be classified under three main headings:

1. *Stress boundary conditions* wherein the stress vector is prescribed at each point on the bounding curve.
2. *Displacement boundary conditions* wherein the displacement vector is prescribed at each point on the bounding curve.
3. *Mixed boundary conditions* wherein the prescription of displacements complements the prescription of stresses in the specification of conditions on the bounding curve: thus the stress vector may be prescribed over part of the boundary and the displacement vector is then prescribed over the complement; alternatively, at any point of the bounding curve may be prescribed one component of the stress vector together with the complementary component of the displacement vector.

In the course of our analysis, we shall obtain a Legendre series representation for each of the stress and displacement components: the coefficients appearing in the representations for the displacements components are expressed in terms of the coefficients occurring in the series for the stress components. From this it will be evident that the treatment of problems associated with either the displacement or the mixed boundary conditions is a straightforward extension of the procedure developed for the problem associated with the stress boundary conditions. These extensions would involve some modifications of the algebra leading to the final determination of the arbitrary constants that arise. Hence, without loss of generality, we confine our attention to a consideration of the stress boundary value problem.

From the standpoint of beam theory, the primary concern is the investigation of the effects in the interior resulting from the conditions applied at the edges: we shall, therefore, give priority to a detailed consideration of the influence of the applied edge stresses. Hence, in our designation of the problem the primary specification consists of the prescription of the stresses along the edges: the prescription of the stresses along the faces together with the geometrical and physcial features of the problem are the secondary specifications. In keeping with the priority stated above, it is appropriate to allow the most general form in the primary specifications.

If one were to aim for maximum generality in every detail, one would also admit the most general form for the secondary features. The retention of such generality would not involve any essential difficulty for the method of analysis: however, it would make the algebra unjustifiably cumbersome and tend to obscure the primary aim, namely, the manifestation of the interior effects due to the applied edge stresses. Accordingly, in the course of the development we shall contain the algebra within the limits necessary for clarity by restricting to relatively simple cases some of the secondary specifications. These restrictions are introduced in successive stages.

In the equilibrium equations of elasticity the material density and the field intensity appear only in the combination formed from the multiplication of the intensity vector by the scalar density; this vector is therefore the body force per unit volume and we refer to it as the body force density. In dealing with the equilibrium equations, we shall restrict ourselves to the case in which the body force density has no axial component and whose transverse component has no transverse variation: (this includes the standard case of a horizontal beam, whose density does not vary through the thickness, subject to a vertical gravity field). In the representation for the stress components, which we then derive from the equilibrium equations, there is no restriction on the *axial* variation of the transverse body force density. The analysis for a completely arbitrary body force density follows the same pattern and for that case the formulae for the stress components are given without derivation.

When we come to an analysis of the constitutive relations, some further secondary features appear – namely, the elastic parameters characterizing the material behavior. With a view to moderating the length of the ensuing algebraic expressions, it is convenient at that point to introduce some further modifications in the secondary specifications. We therefore, confine our attention to those cases where both elastic parameters – namely, Young's modulus and Poisson's ratio – as well as the body force density are constant, but we retain the variability of the beam thickness. Although the subsequent analysis is carried out only for this restricted class of problems, the consideration of any excluded effects could be incorporated by an appropriate extension of the procedure.

Consistent with admitting full generality in the primary specifications we allow an arbitrary distribution of the applied stresses along the edges, assuming only that the distributions are analytic functions of the transverse coordinate. These applied edge stress distributions can be separated into two distinct parts – a principal part characterized by the fact that it contributes to the net edge stress resultants and stress couples, and a subsidiary part consisting of self-equilibrating stress distributions. The principal part is responsible for the principal effects in the interior, the analysis of which is the aim of classical beam theory: on the other hand the latter theory is insensitive to the residual effects induced by the subsidiary part of the edge stress distributions. We further recall that classical beam theory consists of the independent problems of stretching and bending: the corresponding distinction in the general problem appears in the uncoupling of the effects associated respectively with the even and odd parts of the axial normal stress in its dependence on the transverse coordinate.

While physical considerations indicate unambiguously what the principal and residual stress effects should be there is no explicit suggestion for the corresponding displacement effects. The particular method of series expansion employed leads to a formulation of the governing equations in a form in which the mean displacements are the only displacement quantities that appear explicitly. Moreover, there is a partial uncoupling in that the equations for the principal stresses can be immediately integrated, which in turn admits the separated formulation of the equations for the residual stresses. There then remains the equations for the mean displacements: an inspection of the latter equations suggests the manner in which the principal and residual components of the mean displacements should be defined. This resolution completes the uncoupling and leads to two independent systems of equations respectively describing the principal and residual effects. Each system has associated with it its own particular set of terminal conditions thus permitting either problem (the problem for the principal effects or the problem for the residual effects) to be solved without reference to the other.

The objectives of the treatment may now be stated under three main headings:

1. The clarification of the distinction between the principal and residual effects.
2. A systematic derivation of the boundary value problems of classical beam theory (principal effects).
3. The formulation and analysis of the boundary value problems for the residual effects.

The first two sections consist of a statement of the problem together with its reformulation in terms of a new coordinate system, transformed so as to make the region rectangular, and thus facilitating the subsequent variable separation. The third section is concerned with the partial integration of the equilibrium equations leading to Legendre series representations for the stress components. These representations require that the boundary conditions be restated in terms of the coefficients appearing in these series. This is done, for the face boundary conditions, in section four and for the edge boundary conditions in section five. From the reformulation of the conditions on the faces there emerges the stress equilibrium equations of classical beam theory, which are integrated prior to the application of the edge boundary conditions. This leads to the solution of the stress problem of classical beam theory as well as to the conditions for overall static equilibrium.

The content of sections six and seven is an analysis of the constitutive relations which yields the Legendre series representations for the displacement components and also leads to the system of equations to be satisfied by the residual stresses and the mean displacements. By an appropriate separation of the principal and residual effects each of the equations for the mean displacements can be decomposed into an equation for the principal part associated with classical beam theory, and an equation for the residual mean displacement associated with the equations for the residual stresses: this is done in section eight.

There remains the problem of solving for the residual effects. In order to clarify the procedure to be followed, we thereafter restrict our discussion to the case of constant thickness, which results in a considerable simplification. The problem of stretching for this restricted case is considered in section nine and we find that the calculation of the residual mean displacement follows immediately once the residual stresses are known. For the determination of the latter, we have an infinite set of fourth order ordinary differential equations with constant coefficients. In its vector space formulation the solution of such a system amounts to specifying the prescription for the construction of the fundamental matrices. This interesting application of the theory of linear operators in function space concludes the analysis of the stretching problem.

The corresponding treatment for the problem of bending is developed in section ten.

In the typical beam, the characteristic feature is the fact that the axial length is significantly greater than the maximum transverse thickness. The subsequent analysis is carried out without any reference to this mensural characteristic and its validity is not confined to a restricted range of the length-thickness ratio. However, in the analysis of the residual stresses the parameter representing the latter ratio emerges naturally in the solution. Inspection of the solution in its dependence on this parameter then puts the essential features in focus. It will be evident how this ratio effects the rapidity with which the residual effects decay away from the edges and also the extent to which these latter effects are dominated by the principal effects throughout the medium.

We refer to the transverse straightline through the midpoint of the axis as the midsection; the centerline together with the midsection constitute the obvious set of coordinate axes for the problem. The idealized case of a semi-infinite beam having but one edge requires special attention: for this case the centerline together with the edge would be the natural coordinate system. This special case can therefore be conveniently handled as a limiting case of a finite beam after first making appropriate translation of the origin. In previous treatments of the residual problem this idealized case has received the most attention. At the end of the chapter there is a section devoted to a brief discussion of some of the prior work, together with citations to the literature relevant to the problems treated.

1. The Boundary Value Problem

Referred to the orthogonal Cartesian $x-z$ coordinates system in which the x-axis is the centerline and the z-axis is the midsection, the region $\mathcal{R}$ occupied by the figure is defined by

$$\mathcal{R}: \begin{array}{c} -a \leq x \leq a \\ -h(x) \leq z \leq h(x) \end{array} \tag{1.1}$$

Hence, the length is given by the constant $2a$ while the thickness, denoted by $2h(x)$, may vary with the axial coordinate. The bounding curve consists of the pair of edges

$$x = \mp a \tag{1.2a}$$

together with the pair of faces

$$z = \mp h(x) \tag{1.2b}$$

Employing the notation of ordered pairs for two-component vectors, we let $[\tau_{xx}(x,z), \tau_{xz}(x,z)]$ and $[\tau_{zx}(x,z), \tau_{zz}(x,z)]$ denote the stress vectors on elements normal to the x and z axes, respectively. We suppose the medium subject to a transverse body force density, denoted vectorially by $[0, f_0(x)]$, where as indicated, f_0 does not vary through the thickness. Then, using the comma notation for partial differentiation the Cauchy equilibrium conditions consist of the pair of differential equations

$$\tau_{xx,x} + \tau_{xz,z} = 0 \tag{1.3a}$$

$$\tau_{zx,x} + \tau_{zz,z} + f_0 = 0 \tag{1.3b}$$

together with the symmetry relation

$$\tau_{xz} = \tau_{zx} \tag{1.3c}$$

If we denote the displacement vector by $[v_x(x,z),\ v_z(x,z)]$, then, for an isotropic medium, the system of Cauchy-Green constitutive relations, with linearized strain measures, may be written

$$\gamma_{xx} = v_{x,x} = \frac{1}{E}[\tau_{xx} - \nu\tau_{zz}] \tag{1.4a}$$

$$\gamma_{xz} = \frac{1}{2}(v_{x,z} + v_{z,x}) = \frac{1}{E}(1 + \nu)\,\tau_{xz} \tag{1.4b}$$

$$\gamma_{zz} = v_{z,z} = \frac{1}{E}[\tau_{zz} - \nu\tau_{xx}] \tag{1.4c}$$

The above relations define the linearized strain measures, γ_{xx}, γ_{xz} and γ_{zz}, in terms of the displacement components, and also give the relationships between the strain measures and the stress components. The coefficients E and ν are Young's modulus and Poisson's ratio, respectively, for the medium.

Also associated with the displacement quantities is the infinitesimal rotation ω, defined by

$$\omega = \frac{1}{2}(v_{x,z} - v_{z,x}) \tag{1.5}$$

The introduction of the formula for $v_{x,z}$ from (1.4b) into (1.5) gives an alternative form for the rotation quantity, namely

$$\omega = \frac{1}{E}(1 + \nu)\tau_{xz} - v_{z,x} \tag{1.6}$$

We consider the lower and upper faces of the beam subject to arbitrary applied surface stress vectors which we denote by $[p_x^-(x),\ p_z^-(x)]$ and $[p_x^+(x),\ p_z^+(x)]$, respectively: in these we have used the x-coordinate as the face parameter. If we denote dh/dx by h', the boundary conditions on the lower face take the form

$$z = -h: \quad \left\{ \begin{aligned} &\frac{1}{\sqrt{1 + h'^2}}\,[\tau_{xz}(x, -h) + h'\tau_{xx}(x, -h)] = p_x^- &\text{(1.7a)} \\ &\frac{1}{\sqrt{1 + h'^2}}\,[\tau_{zz}(x, -h) + h'\tau_{zx}(x, -h)] = p_z^- &\text{(1.7b)} \end{aligned} \right.$$

while the corresponding conditions on the upper face read

$$z = h: \quad \begin{cases} \dfrac{1}{\sqrt{1 + h'^2}} [\tau_{xz}(x,h) - h'\tau_{xx}(x,h)] = p_x^+ & \textbf{(1.8a)} \\[2ex] \dfrac{1}{\sqrt{1 + h'^2}} [\tau_{zz}(x,h) - h'\tau_{zx}(x,h)] = p_z^+ & \textbf{(1.8b)} \end{cases}$$

Along the left and right edges there are applied the stress vectors $[\tau_{xx}^-(z), \tau_{xz}^-(z)]$ and $[\tau_{xx}^+(z), \tau_{xz}^+(z)]$, respectively, where the four functions $\tau_{xx}^{\mp}(z)$ and $\tau_{xz}^{\mp}(z)$ are arbitrary functions of z: hence for the edge boundary conditions, we have

$$x = -a: \ \tau_{xx}(-a,z) = \tau_{xx}^-(z), \ \tau_{xz}(-a,z) = \tau_{xz}^-(z) \qquad \textbf{(1.9a)}$$

$$x = +a: \ \tau_{xx}(a,z) = \tau_{xx}^+(z), \ \tau_{xz}(a,z) = \tau_{xz}^+(z) \qquad \textbf{(1.9b)}$$

The stress boundary value problem for a two-dimensional elastic medium is that of determining the unknown field quantities, namely the stress and displacement components, throughout the region $\mathcal{R}$. This consists in solving, explicitly in terms of x and z, the system of partial differential equations (1.3) and (1.4) subject to the boundary conditions (1.7) to (1.9).

2. Normalization of the Transverse Coordinate

We introduce a coordinate transformation that leaves the x coordinate unchanged and replaces the z coordinate by the normalized transverse variable t, defined by

$$t = \frac{z}{h(x)} \qquad \textbf{(2.1)}$$

from which it immediately follows that

$$\frac{\partial t}{\partial x} = -\frac{h'}{h}t, \quad \frac{\partial t}{\partial z} = \frac{1}{h} \qquad \textbf{(2.2)}$$

Under the transformation (2.1), the region $\mathcal{R}$ in the $x-z$ plane defined by (1.1) becomes the rectangle $\mathcal{R}_0$ in the $x-t$ plane defined by

$$\mathcal{R}_0: \quad \begin{matrix} -a \le x \le a \\ -1 \le t \le 1 \end{matrix} \qquad \textbf{(2.3)}$$

The rectilinear boundary now consists of the pair of edges

$$x = \mp a \qquad \textbf{(2.4a)}$$

together with the pair of faces

$$t = \mp 1 \tag{2.4b}$$

In reformulating the boundary value problem for the region $\mathcal{R}_0$, all field quantities are to be considered as functions of x and t. To emphasize the transformed dependence, we introduce new notation that reflects the effect of the transformation (2.1). For the stress quantities, we set

$$\sigma_x(x,t) = \tau_{xx}(x,z), \quad \sigma(x,t) = \tau_{zz}(x,z) \tag{2.5a}$$

$$\tau(x,t) = \tau_{xz}(x,z) = \tau_{zx}(x,z) \tag{2.5b}$$

and for the edge distributions, we write

$$\sigma_x^{\mp}(t) = \tau_{xx}^{\mp}(z), \quad \tau^{\mp}(t) = \tau_{xz}^{\mp}(z) \tag{2.5c}$$

Similarly for the displacement vector, we define the components u and w by

$$u(x,t) = v_x(x,z), \quad w(x,t) = v_z(x,z) \tag{2.6}$$

Applying the rules for partial differentiation and using relations (2.2), it follows that

$$\begin{aligned} \tau_{xx,x} &= \sigma_{x,x} - \frac{h'}{h}\, t\, \sigma_{x,t}, & \tau_{xz,z} &= \frac{1}{h}\tau,_t \\ \tau_{zx,x} &= \tau,_x - \frac{h'}{h}\, t\, \tau,_t, & \tau_{zz,z} &= \frac{1}{h}\sigma,_t \end{aligned} \tag{2.7}$$

and furthermore

$$\begin{aligned} v_{x,x} &= u,_x - \frac{h'}{h}\, t\, u,_t, & v_{x,z} &= \frac{1}{h} u,_t \\ v_{z,x} &= w,_x - \frac{h'}{h}\, t\, w,_t, & v_{z,z} &= \frac{1}{h} w,_t \end{aligned} \tag{2.8}$$

Relations (2.7) and (2.8) are now to be introduced into the equations of Section 1.

Under the above transformations, the equilibrium equations (1.3) take the form

$$h\sigma_{x,x} - h'\,t\,\sigma_{x,t} + \tau,_t = 0 \tag{2.9a}$$

$$h\tau,_x - h'\,t\,\tau,_t + \sigma,_t + hf_0 = 0 \tag{2.9b}$$

while the stress-displacement relations (1.4) become

$$hu_{,x} - h't u_{,t} = \frac{h}{E}[\sigma_x - \nu\sigma] \qquad (2.10a)$$

$$u_{,t} + hw_{,x} - h't w_{,t} = 2\frac{h}{E}(1 + \nu)\tau \qquad (2.10b)$$

$$w_{,t} = \frac{h}{E}[\sigma - \nu\sigma_x] \qquad (2.10c)$$

and the rotation formula (1.6) reads

$$\omega = \frac{1}{E}(1 + \nu)\tau - w_{,x} + \frac{h'}{h}t w_{,t} \qquad (2.11)$$

Corresponding to conditions (1.7) to (1.9), we now have the transformed face conditions

$$t = -1: \begin{cases} \dfrac{1}{\sqrt{1 + h'^2}}[\tau(x, -1) + h'\sigma_x(x, -1)] = p_x^- & (2.12a) \\[4em] \dfrac{1}{\sqrt{1 + h'^2}}[\sigma(x, -1) + h'\tau(x, -1)] = p_z^- & (2.12b) \end{cases}$$

and

$$t = 1: \begin{cases} \dfrac{1}{\sqrt{1 + h'^2}}[\tau(x, 1) - h'\sigma_x(x, 1)] = p_x^+ & (2.13a) \\[4em] \dfrac{1}{\sqrt{1 + h'^2}}[\sigma(x, 1) - h'\tau(x, 1)] = p_z^+ & (2.13b) \end{cases}$$

together with the transformed edge conditions

$$x = -a: \quad \sigma_x(-a, t) = \sigma^-(t), \quad \tau(-a, t) = \tau^-(t) \qquad (2.14a)$$

$$x = +a: \quad \sigma_x(a, t) = \sigma^+(t), \quad \tau(a, t) = \tau^+(t) \qquad (2.14b)$$

The transformed boundary value problem consists of determining the solution to the system of equations (2.9) to (2.11), valid in the region $\Re_0$, and satisfying the set of boundary conditions (2.12) to (2.14).

3. Representations for the Stress Components

In obtaining representations for the stress components, our procedure involves taking an explicit form for the dependence on t leaving the implicit dependence on x to be determined later. Assuming only that the stress components depend analytically on the normalized transverse coordinate, we start by expanding the axial normal stress in a series of Legendre polynomials in t where the coefficients are unknown functions of x: then, by integration of the equilibrium equations, we derive the corresponding Legendre expansions for the other two stress components.

This procedure yields representations for the stresses in separated form; namely, the dependence appears in the form of linear combinations of terms, each of which is the product of a function of x by a function of t. In that form, derivatives appear only as total derivatives for which it is convenient to have an abbreviated notation: we shall use a dot to denote d/dt and, as before, a prime to denote d/dx.

The integration of the Legendre series is facilitated by the use of the recurrence relations for Legendre polynomials. Letting $P_n(t)$ denote the Legendre polynomial of degree n, we have*

$$(2n+1)\,P_n(t) = \dot{P}_{n+1}(t) - \dot{P}_{n-1}(t) \tag{3.1}$$

from which it follows that apart from an additive term independent of t, we may write

$$\int P_n(t)\,dt = \frac{1}{2n+1}\left[P_{n+1}(t) - P_{n-1}(t)\right] \tag{3.2}$$

A second recurrence relation, namely

$$t\dot{P}_n(t) = nP_n(t) + \dot{P}_{n-1}(t) \tag{3.3}$$

expedites the integration of those terms arising from the inclusion of effects due to variation in the thickness.

We write the expansion for σ_x in the form

$$\sigma_x = \sum_{n=0}^{\infty} s_n(x)\,P_n(t) \tag{3.4}$$

in which the coefficients s_n are undetermined functions of x: it immediately follows that

$$\sigma_{x,x} = \sum_{n=0}^{\infty} s_n'(x)\,P_n(t), \quad \sigma_{x,t} = \sum_{n=1}^{\infty} s_n(x)\,\dot{P}_n(t) \tag{3.5}$$

where, in the second relation, we have noted that $\dot{P}_0 = 0$. If we introduce relations (3.5) into the first equilibrium equation (2.9a), we obtain the equation for τ, namely

$$\tau,_t = -h\sum_{n=0}^{\infty} s_n' P_n + h'\sum_{n=1}^{\infty} s_n\,t\dot{P}_n \tag{3.6}$$

* For a derivation of the recurrence formulae (3.1), (3.3) and (6.9) for the Legendre polynomials, we refer to § 15.21 of "Modern Analysis" by E. T. Whittaker and G. N. Watson, C.U.P., Cambridge, 1952.

where we no longer write the dependence on x and t explicitly. Applying the recurrence relation (3.3) to the terms in the second summation on the right of (3.6), we have

$$\tau,_t = -h \sum_{n=0}^{\infty} s_n' P_n + h' \sum_{n=1}^{\infty} s_n (nP_n + \dot{P}_{n-1}) \tag{3.7}$$

which on rearrangement reads

$$\tau,_t = - \sum_{n=0}^{\infty} (hs_n' - nh's_n)P_n + h' \sum_{n=1}^{\infty} s_{n+1}\dot{P}_n \tag{3.8}$$

where again we have noted that $\dot{P}_0 = 0$.

The integration of the above equation with respect to t yields the Legendre series representation for τ. The integration of the second summation is straightforward: to integrate the first summation, we use formula (3.2) and rearrange the terms. We obtain

$$\tau = \sum_{n=0}^{\infty} \tau_n(x)\ P_n(t) \tag{3.9}$$

in which τ_0 is an undetermined function of x introduced by the integration, and the higher coefficients are given by

$$\tau_n = h\left[\frac{s_{n+1}'}{2n+3} - \frac{s_{n-1}'}{2n-1}\right] - h'\left[\frac{n+1}{2n+3}s_{n+1} - \frac{n-1}{2n-1}s_{n-1}\right] + h's_{n+1},\quad n \geq 1 \tag{3.10}$$

$$= h\left[\frac{s_{n+1}'}{2n+3} - \frac{s_{n-1}'}{2n-1}\right] + h'\left[\frac{n+2}{2n+3}s_{n+1} + \frac{n-1}{2n-1}s_{n-1}\right],\qquad n \geq 1. \tag{3.10*}$$

The corresponding representation for the transverse normal stress is obtained in an identical manner from the second equilibrium equation. Introducing the expansion (3.9) into (2.9b) and using the recurrence relation (3.3), we rearrange the terms to obtain

$$\sigma,_t = - \sum_{n=0}^{\infty} (h\tau_n' - nh'\tau_n)P_n + h' \sum_{n=0}^{\infty} \tau_{n+1}\dot{P}_n - hf_0 \tag{3.11}$$

the integration of which gives the Legendre series representation for σ: we find

$$\sigma = - \sum_{n=0}^{\infty} \sigma_n(x)\ P_n(t) - h(x)\, f_0(x)\ P_1(t) \tag{3.12}$$

in which σ_0 is an undetermined function of x introduced by the integration and the higher coefficients are given by

$$\sigma_n = h\left[\frac{\tau'_{n+1}}{2n+3} - \frac{\tau'_{n-1}}{2n-1}\right] - h'\left[\frac{n+1}{2n+3}\tau_{n+1} - \frac{n-1}{2n-1}\tau_{n-1}\right] + h'\tau_{n+1}, \quad n \geq 1 \tag{3.13}$$

$$= h\left[\frac{\tau'_{n+1}}{2n+3} - \frac{\tau'_{n-1}}{2n-1}\right] + h'\left[\frac{n+2}{2n+3}\tau_{n+1} + \frac{n-1}{2n-1}\tau_{n-1}\right], \qquad n \geq 1. \tag{3.13*}$$

In fact, if we substitute for the τ_n in terms of the s_n from (3.10), and rearrange, we obtain for $n \geq 2$.

$$\begin{aligned}\sigma_n = {} & h^2\left[\frac{s''_{n-2}}{(2n-3)(2n-1)} - \frac{2s''_n}{(2n-1)(2n+3)} + \frac{s''_{n+2}}{(2n+3)(2n+5)}\right] \\ & - hh'\left[\frac{2(n-2)s'_{n-2}}{(2n-3)(2n-1)} + \frac{6s'_n}{(2n-1)(2n+3)} - \frac{2(n+3)s'_{n+2}}{(2n+3)(2n+5)}\right] \\ & + h'^2\left[\frac{(n-2)(n-1)s_{n-2}}{(2n-3)(2n-1)} + \frac{(2n^2+2n-3)s_n}{(2n-1)(2n+3)} + \frac{(n+2)(n+3)s_{n+2}}{(2n+3)(2n+5)}\right] \\ & - hh''\left[\frac{(n-2)s_{n-2}}{(2n-3)(2n-1)} + \frac{3s_n}{(2n-1)(2n+3)} - \frac{(n+3)s_{n+2}}{(2n+3)(2n+5)}\right], \qquad n \geq 2. \end{aligned} \tag{3.14}$$

The expansions (3.4), (3.9) and (3.12), together with formulae (3.10) and (3.13), give the Legendre expansions for the three stress components: besides the original unknown functions, $\{s_n(x) : n \geq 0\}$, they also involve the undetermined quantities $\tau_0(x)$ and $\sigma_0(x)$. As the stress field is thus entirely represented, so the solution of the boundary value problem lies in the determination of these coefficients.

We shall see presently that the satisfaction of the face boundary conditions leads to the equations to be satisfied by the quantities s_0, s_1, τ_0 and σ_0: the equations for the remaining unknowns, $\{s_n(x),\ n \geq 2\}$, will follow from the utilization of the stress-displacement relations. The edge boundary conditions then give the side conditions determining the unique solution of these equations.

Addendum: We conclude this section with a statement of the corresponding results for the case of a general body force density. Instead of the vector $[0, f_0]$ we envision the general form $[f^x, f]$, where both f^x and f are arbitrary analytic functions of x and t, and write the Legendre expansions of f^x and f in the form

$$f^x = \sum_{n=0}^{\infty} f_n^x(x)\, P_n(t), \quad f = \sum_{n=0}^{\infty} f_n(x)\, P_n(t)$$

If we use a tilde to distinguish the coefficients in this case, then, corresponding to (3.4), we assume an expansion for the axial normal stress in the form

$$\sigma_x = \sum_{n=0}^{\infty} \tilde{s}_n(x)\, P_n(t)$$

and the expansions corresponding to (3.9) and (3.12) for the other two stress components take the form

$$\tau = \sum_{n=0}^{\infty} \tilde{\tau}_n(x)\, P_n(t), \quad \sigma = \sum_{n=0}^{\infty} \tilde{\sigma}_n(x)\, P_n(t)$$

where $\tilde{\tau}_0$ and $\tilde{\sigma}_n$ are undetermined and the remaining coefficients are given by

$$\tilde{\tau}_n = h\left[\frac{\tilde{s}_{n+1}}{2n+3} - \frac{\tilde{s}_{n-1}}{2n-1}\right] + h'\left[\frac{n+2}{2n+3}\tilde{s}_{n+1} + \frac{n-1}{2n-1}\tilde{s}_{n-1}\right]$$
$$+ h\left[\frac{f^x_{n+1}}{2n+3} - \frac{f^x_{n-1}}{2n-1}\right],$$

$$\tilde{\sigma}_n = h\left[\frac{\tilde{\tau}_{n+1}}{2n+3} - \frac{\tilde{\tau}_{n-1}}{2n-1}\right] + h'\left[\frac{n+2}{2n+3}\tilde{\tau}_{n+1} + \frac{n-1}{2n-1}\tilde{\tau}_{n-1}\right]$$
$$+ h\left[\frac{f_{n+1}}{2n+3} - \frac{f_{n-1}}{2n-1}\right].$$

The above are listed merely to indicate how the effects of the general body force density can be incorporated: we shall not refer to them further in the sequel.

4. The Face Boundary Conditions

The application of the two pairs of boundary conditions for the lower and upper faces leads to four relations between the coefficients appearing in the representations for the stresses. By appropriate combinations of these relations, we arrive at a simpler but equivalent set which are respectively

1) the differential equation to be satisfied by s_0,
2) the differential equation to be satisfied by s_1,
3) the formula for τ_0 in terms of s_1,
4) the formula for σ_0 in terms of s_0 and s_2.

In fact we shall see that, ignoring the last relation, the remaining three can be interpreted as the equilibrium equations of classical beam theory.

Recalling that for Legendre polynomials,

$$P_n(-1) = (-1)^n,\ P_n(1) = 1 \tag{4.1}$$

it immediately follows that

$$\sigma_x(x,-1) = s_0 + \sum_{n=1}^{\infty} (-1)^n s_n,\ \sigma_x(x,+1) = s_0 + \sum_{n=1}^{\infty} s_n \tag{4.2}$$

$$\tau(x,-1) = \tau_0 + \sum_{n=1}^{\infty} (-1)^n \tau_n,\ \tau(x,+1) = \tau_0 + \sum_{n=1}^{\infty} \tau_n \tag{4.3}$$

$$\sigma(x,-1) = \sigma_0 + \sum_{n=1}^{\infty} (-1)^n \sigma_n,\ \sigma(x,+1) = \sigma_0 + \sum_{n=1}^{\infty} \sigma_n \tag{4.4}$$

We next insert formula (3.10) for the τ_n into relations (4.3): if we note the cancellations resulting from the recursive features of (3.10), we obtain

$$\tau(x,-1) = \tau_0 + hs_0' - \frac{1}{3}hs_1' + \frac{1}{3}h's_1 + h'\sum_{n=2}^{\infty}(-1)^{n-1}s_n \tag{4.5a}$$

$$\tau(x,+1) = \tau_0 - hs_0' - \frac{1}{3}hs_1' + \frac{1}{3}h's_1 + h'\sum_{n=2}^{\infty}s_n \tag{4.5b}$$

Similarly by introducing formula (3.13) for the σ_n into (4.4), and again making the obvious cancellations, there results

$$\sigma(x,-1) = \sigma_0 + h\tau_0' - \frac{1}{3}h\tau_1' + \frac{1}{3}h'\tau_1 + h'\sum_{n=2}^{\infty}(-1)^n\tau_n + hf_0 \tag{4.6a}$$

$$\sigma(x,+1) = \sigma_0 - h\tau_0' - \frac{1}{3}h\tau_1' + \frac{1}{3}h'\tau_1 + h'\sum_{n=2}^{\infty}\tau_n - hf_0 \tag{4.6b}$$

An appropriate combination of (4.2) with (4.5) gives

$$\tau(x,-1) + h'\sigma_x(x,-1) = \tau_0 + hs_0' + h's_0 - \frac{1}{3}hs_1' - \frac{2}{3}h's_1 \tag{4.7a}$$

$$\tau(x,+1) - h'\sigma_x(x,+1) = \tau_0 - hs_0' - h's_0 - \frac{1}{3}hs_1' - \frac{2}{3}h's_1 \tag{4.7b}$$

while a similar combination of (4.3) with (4.6) yields

$$\sigma(x,-1) + h'\tau(x,-1) = \sigma_0 + h\tau_0' + h'\tau_0 - \frac{1}{3}h\tau_1' - \frac{2}{3}h'\tau_1 + hf_0 \tag{4.8a}$$

$$\sigma(x,+1) - h'\tau(x,-1) = \sigma_0 - h\tau_0' - h'\tau_0 - \frac{1}{3}h\tau_1' - \frac{2}{3}h'\tau_1 - hf_0 \tag{4.8b}$$

By means of relations (4.7), the boundary conditions (2.12a) and (2.13a) become, respectively

$$\tau_0 + (hs_0)' - \frac{1}{3h}(h^2 s_1)' = \sqrt{1 + h'^2}\,\overset{-}{p}_x \tag{4.9a}$$

$$\tau_0 - (hs_0)' - \frac{1}{3h}(h^2 s_1)' = \sqrt{1 + h'^2}\,\overset{+}{p}_x \tag{4.9b}$$

while the use of relations (4.8) renders the boundary conditions (2.12b) and (2.13b), respectively, in the form

$$\sigma_0 + (h\tau_0)' - \frac{1}{3h}(h^2\tau_1)' = \sqrt{1 + h'^2}\,\overset{-}{p}_z - hf_0 \tag{4.10a}$$

$$\sigma_0 - (h\tau_0)' - \frac{1}{3h}(h^2\tau_1)' = \sqrt{1 + h'^2}\,\overset{+}{p}_z + hf_0 \tag{4.10b}$$

If we introduce the notation

$$\overset{+}{p}_x + \overset{-}{p}_x = p_x^*, \quad \overset{+}{p}_x - \overset{-}{p}_x = p_x \tag{4.11a}$$

$$\overset{+}{p}_z + \overset{-}{p}_z = p^*, \quad \overset{+}{p}_z - \overset{-}{p}_z = p \tag{4.11b}$$

then subtraction and addition of equations (4.9) yield, respectively

$$(2hs_0)' + \sqrt{1 + h'^2}\,p_x = 0 \tag{4.12}$$

and

$$(\tfrac{2}{3} h^2 s_1)' - 2h\tau_0 + h\sqrt{1 + h'^2}\, p_x^* = 0 \tag{4.13}$$

while similar operations on equations (4.10) give

$$(2h\tau_0)' + \sqrt{1 + h'^2}\, p + 2hf_0 = 0 \tag{4.14}$$

and

$$(\tfrac{2}{3} h^2 \tau_1)' - 2h\sigma_0 + h\sqrt{1 + h'^2}\, p^* = 0 \tag{4.15}$$

respectively. Substituting for τ_0 from (4.13) into (4.14), we obtain

$$(\tfrac{2}{3} h^2 s_1)'' + \sqrt{1 + h'^2}\, p + 2hf_0 + (h\sqrt{1 + h'^2}\, p_x^*)' = 0 \tag{4.16}$$

Relations (4.12) and (4.16) are the differential equations to be satisfied by s_0 and s_1, respectively, while τ_0 must satisfy (4.14).

We may rewrite equation (4.13) in the form

$$\tau_0 = \frac{1}{3h}(h^2 s_1)' + \frac{1}{2}\sqrt{1 + h'^2}\, p_x^* \tag{4.17}$$

If we recall the system (3.10*), it is clear that formula (4.17) is the relation required for the excluded case $n = 0$. The system (3.10*) complemented by (4.17) is the complete set of relations giving the τ_n in terms of the s_n.

Similarly, if we write equation (4.15) in the form

$$\sigma_0 = \frac{1}{3h}(h^2 \tau_1)' + \frac{1}{2}\sqrt{1 + h'^2}\, p^* \tag{4.18a}$$

we have the relation required for the excluded case $n = 0$ in (3.13*). It is convenient for our purpose to note the corresponding formula for σ_1, namely

$$\sigma_1 = \frac{1}{5}(h\tau_2' + 3h'\tau_2) - h\tau_0' \tag{4.18b}$$

obtained by setting $n = 1$ in (3.13*).

An appropriate substitution transforms relations (4.18) into formulae expressing σ_0 and σ_1 in terms of the s_n, thus providing for the cases excluded from (3.14). For this we note the formulae for τ_1 and τ_2 implied by (3.10*), namely

$$\tau_1 = \frac{1}{5}(hs_2' + 3h's_2) - hs_0' \tag{4.19a}$$

$$\tau_2 = \frac{1}{7}(hs_3' + 4h's_3) - \frac{1}{3}(hs_1' - h's_1) \tag{4.19b}$$

We introduce (4.19a) into (4.18a) and use relations (4.17) and (4.19b) in (4.18b): after some rearrangement, we obtain

$$\sigma_0 = \frac{1}{15h}(h^3 s_2)'' - \frac{1}{3h}(h^3 s_0')' + \frac{1}{2}\sqrt{1 + h'^2}\, p^* \tag{4.20a}$$

$$\sigma_1 = \frac{1}{35}[h^2 s_3'' + 8hh' s_3' + 4(hh'' + 3h'^2)s_3]$$

$$- \frac{1}{5}[2h^2 s_1'' + 6hh' s_1' + (3hh'' - h'^2)s_1 + \frac{5h}{2}(\sqrt{1 + h'^2}\, p_x^*)'] \tag{4.20b}$$

The system (3.14) complemented by relations (4.20) constitute the complete set of relations expressing the σ_n in terms of the s_n. We can obtain an alternative form for (4.20a) if we use (4.12) to express s_0' in terms of s_0: explicitly, we obtain the equivalent expression

$$\sigma_0 = \frac{1}{15h}(h^3 s_2)'' + \frac{1}{3h}(h^2 h' s_0)' + \frac{1}{2}\sqrt{1 + h'^2}\, p^* + \frac{1}{6h}[h^2\sqrt{1 + h'^2}\, p_x]' \tag{4.20a*}$$

if we note from (4.12) that

$$hs_0' = -h' s_0 - \frac{1}{2}\sqrt{1 + h'^2}\, p_x\,. \tag{4.21}$$

An inspection of the above relations reveals the immediate effects of the surface and body forces on the various stress components. The body force and three of the surface resultants directly induce axial stresses. Equation (4.12) shows the effect of p_x on the coefficient s_0 while equation (4.16) indicates the influence of the quantities f_0, p and p_x^* on the coefficient s_1. Consistent with the classification introduced later, the surface resultant p_x is a stretching effect while the body force f_0, and the net transverse pressure p are bending effects as is also the quantity p_x^*. The appearance of the latter term in formula (4.17) for τ_0 shows that it also directly induces a transverse shear stress: its related effect on the transverse normal stress is seen in formula (4.20b) for σ_1. In formulae (4.18a) and (4.20a) for σ_0, there occurs the remaining surface quantity p^*, also associated with the stretching effects: if we note that p^* measures the "pinching" tendency of the surface forces, we see that this direct influence on the transverse normal stress is to be expected.

For future reference we also note here the explicit form for σ_2 and σ_3 given by (3.14), namely

$$\sigma_2 = h^2\left[\frac{1}{3}s_0'' - \frac{2}{21}s_2'' + \frac{1}{63}s_4''\right] - hh'\left[\frac{2}{7}s_2' - \frac{10}{63}s_4'\right]$$

$$+ h'^2\left[\frac{3}{7}s_2 + \frac{20}{63}s_4\right] - hh''\left[\frac{1}{7}s_2 - \frac{5}{63}s_4\right] \tag{4.22a}$$

$$\sigma_3 = h^2 \left[\frac{1}{15} s_1'' - \frac{2}{45} s_3'' + \frac{1}{99} s_5'' \right] - hh' \left[\frac{2}{15} s_1' + \frac{2}{15} s_3' - \frac{4}{33} s_5' \right]$$

$$+ h'^2 \left[\frac{2}{15} s_1 + \frac{7}{15} s_3 + \frac{10}{33} s_5 \right] - hh'' \left[\frac{1}{15} s_1 + \frac{1}{15} s_3 - \frac{2}{33} s_5 \right] \quad \textbf{(4.22b)}$$

Finally, we note that in classical beam theory the stress resultants and couple are defined by

$$N = \int_{-z}^{z} \tau_{xx}\, dz = h \int_{-1}^{1} \sigma_x\, dt = 2hs_0 \quad \textbf{(4.23a)}$$

$$M = \int_{-z}^{z} z\, \tau_{xx}\, dz = h^2 \int_{-1}^{1} \sigma_x\, dt = \frac{2h^2}{3} s_1 \quad \textbf{(4.23b)}$$

$$Q = \int_{-z}^{z} \tau_{xz}\, dz = h \int_{-1}^{1} \tau\, dt = 2h\tau_0 \quad \textbf{(4.23c)}$$

With this notation, equations (4.12) to (4.14) take the familiar form

$$N' + \sqrt{1 + h'^2}\, p_x = 0 \quad \textbf{(4.24)}$$

$$M' - Q + h\sqrt{1 + h'^2}\, p_x^{*} = 0 \quad \textbf{(4.25)}$$

$$Q' + \sqrt{1 + h'^2}\, p + 2hf_0 = 0 \quad \textbf{(4.26)}$$

Thus equations (4.12) to (4.14) are the equilibrium conditions for elementary cross sections of the beam.

5. The Edge Boundary Conditions

Having expressed the face conditions in terms of the coefficients appearing in the Legendre series for the stresses, we now effect a corresponding reformulation of the edge boundary conditions. The edge conditions on the axial normal stress are equivalent to a pair of point boundary conditions on each of the coefficients s_n: similarly the edge conditions on the shear stress are equivalent to a pair of point boundary conditions on each of the coefficients τ_n. Anticipating the later formulation of the equations to be satisfied by the yet undetermined coefficients s_n, $n \geq 2$, we shall transform these latter conditions on the τ_n into a second set of conditions on the s_n.

The face conditions led to the differential equations to be satisfied by s_0, s_1 and τ_0: the determination of these quantities will follow from the integration of the equations together with the application of the edge conditions, from which will also emerge the requirement that certain relations hold between the lower coefficients in the Legendre expansions for the prescribed edge stresses. These latter relations express the constraints necessary for overall static equilibrium.

Considering the right hand sides of relations (2.14), we expand the prescribed stress distributions $\sigma_x^{\mp}(t)$ and $\tau^{\mp}(t)$ in the series of Legendre polynomials as follows

$$\sigma_x^{\mp}(t) = \sum_{n=0}^{\infty} s_n^{\mp} P_n(t), \quad \tau^{\mp}(t) = \sum_{n=0}^{\infty} \tau_n^{\mp} P_n(t) \tag{5.1}$$

where the coefficients $s_n^{\mp}$ and the $\tau_n^{\mp}$ are constant: noting the expansions (3.4) and (3.9), the conditions (2.14) may now be written

$$s_n(\mp a) = s_n^{\mp}, \quad \tau_n(\mp a) = \tau_n^{\mp}, \; n \geq 0 \tag{5.2}$$

In particular we have the conditions

$$s_0(-a) = s_0^{-}, \quad s_0(+a) = s_0^{+} \tag{5.3a}$$

$$s_1(-a) = s_1^{-}, \quad s_1(+a) = s_1^{+} \tag{5.3b}$$

$$\tau_0(-a) = \tau_0^{-}, \quad \tau_0(+a) = \tau_0^{+} \tag{5.3c}$$

associated respectively with equations (4.12), (4.13) and (4.14): the remaining conditions in (5.2) are associated with the system of equations for the remaining s_n, $n \geq 2$ to be formulated later.

In order to transform the second set of conditions in (5.2), namely those on the τ_n, into a form expressing requirements on the s_n, we use relations (3.10) together with relation (4.17) for τ_0. If we write successively the expressions for the τ_n of even index, we obtain

$$\tau_0 = \frac{1}{3} h s_1' + \frac{2}{3} h' s_1 + \frac{1}{2}\sqrt{1 + h'^2}\, p_x^*$$

$$\tau_2 = h\left(\frac{1}{7} s_3' - \frac{1}{3} s_1'\right) - h'\left(\frac{3}{7} s_3 - \frac{1}{3} s_1\right) + h' s_3$$

$$\vdots \qquad\qquad \vdots \qquad\qquad \vdots$$

$$\tau_{2k} = h\left(\frac{s_{2k+1}'}{4k+3} - \frac{s_{2k-1}'}{4k-1}\right) - h'\left(\frac{2k+1}{4k+3} s_{2k+1} - \frac{2k-1}{4k-1} s_{2k-1}\right) + h' s_{2k+1}$$

The addition of the above gives, for $k \geq 0$,

$$\sum_{j=0}^{k} \tau_{2j} = \frac{h s_{2k+1}'}{4k+3} - \frac{2k+1}{4k+3} h' s_{2k+1} + h' \sum_{j=0}^{k} s_{2j+1} + \frac{1}{2}\sqrt{1 + h'^2}\, p_x^* \tag{5.4}$$

from which there follows

$$hs'_{2k+1} = (4k+3)\sum_{j=0}^{k} \tau_{2j} + (2k+1)h' s_{2k+1}$$

$$- (4k+3)\,[h' \sum_{j=0}^{k} s_{2j+1} + \frac{1}{2}\sqrt{1 + h'^2}\; p_x^*],\ k \geq 0\ . \quad \textbf{(5.5)}$$

We can write the expressions for τ_n of odd index and perform a similar summation: if we also use (4.21) to substitute for hs'_0, we obtain for $k \geq 1$.

$$\sum_{j=0}^{k-1} \tau_{2j+1} = \frac{hs'_{2k}}{4k+1} - \frac{2k}{4k+1} h' s_{2k} + h' \sum_{j=0}^{k} s_{2j} + \frac{1}{2}\sqrt{1 + h'^2}\; p_x \quad \textbf{(5.6)}$$

from which we have

$$hs'_{2k} = (4k+1)\sum_{j=0}^{k-1} \tau_{2j+1} + 2kh' s_{2k}$$

$$- (4k+1)\,[h' \sum_{j=0}^{k} s_{2j} + \frac{1}{2}\sqrt{1 + h'^2}\; p_x],\ k \geq 1\ . \quad \textbf{(5.7)}$$

Equations (5.5) may be considered the generalization to higher indices of relation (4.13): the corresponding analog to relation (4.12) is given by equations (5.7). We shall use these relations only at their terminal values to determine edge conditions on the s'_n. We see that if we set

$$d_{2k}^{\mp} = (4k+1)\sum_{j=0}^{k-1} \tau_{2j+1}^{\mp} + 2kh'(\mp a)\, s_{2k}^{\mp}$$

$$- (4k+1)\,[h'(\mp a) \sum_{j=0}^{k} s_{2j}^{\mp} + \frac{1}{2}\sqrt{1 + h'^2(\mp a)}\; p_x(\mp a)],\ k \geq 1\ .$$

(5.8a)

$$d_{2k+1}^{\mp} = (4k+3)\sum_{j=0}^{k} \tau_{2j}^{\mp} + (2k+1)h'(\mp a)\, s_{2k+1}^{\mp}$$

$$- (4k+3)[h'(\mp a) \sum_{j=0}^{k} s_{2j+1}^{\mp} + \frac{1}{2}\sqrt{1 + h'^2(\mp a)}\; p_x^*(\mp a],\ k \geq 0\ .$$

(5.8b)

then the edge conditions (5.2) may be replaced by

$$s_n(\mp a) = s_n^{\mp},\ n \geq 0:\ \ hs_n'(\mp a) = d_n^{\mp},\ n \geq 1. \tag{5.9}$$

which, for convenience in later reference, we separate into the three sets

$$s_0(\mp a) = s_0^{\mp} \tag{5.9a}$$

$$s_1(\mp a) = s_1^{\mp},\ \ hs_1'(\mp a) = d_1^{\mp} \tag{5.9b}$$

$$s_n(\mp a) = s_n^{\mp},\ \ hs_n'(\mp a) = d_n^{\mp},\ n \geq 2\ . \tag{5.9c}$$

The set (5.9a) which is identical with conditions (5.3a) is associated with differential equations (4.12). The second group, namely (5.9b), equivalent to conditions (5.3b) and (5.3c) combined, gives the set of conditions to be assigned to the differential equation (4.16). The third set, (5.9c), constitutes the system of boundary conditions complementing the system of differential equations for the $\{s_n, n > 2\}$ to be formulated in Section Seven.

Since the two conditions (5.9a) are associated with a first order differential equation, they must satisfy a compatibility relation: similarly the four conditions (5.9b) associated with a second order differential equation must satisfy two compatibility relations. These compatibility requirements, giving the conditions for overall static equilibrium, follow from the integration of the differential equations.

Starting with equation (4.12), we integrate over the beam length: if we use the notation

$$\overset{-}{h} = h(-a),\ \ \overset{+}{h} = h(+a) \tag{5.10}$$

and note the edge conditions (5.3a), we are led to the constraint

$$2(\overset{+}{h}\overset{+}{s_0} - \overset{-}{h}\overset{-}{s_0}) = -\int_{-a}^{a} \sqrt{1+h'^2}\ p_x\, dx \tag{5.11}$$

to be satisfied by the coefficients $s_0^{\mp}$. The indefinite integration of equation (4.12) together with the application of the edge conditions and the use of condition (5.11) yields the following alternate forms for the coefficient s_0, namely

$$2hs_0 = 2\overset{-}{h}\ \overset{-}{s_0} - \int_{-a}^{x} \sqrt{1+h'^2}\ p_x\, dx \tag{5.12}$$

$$2hs_0 = \overset{+}{h}\overset{+}{s_0} + \overset{-}{h}\overset{-}{s_0} + \frac{1}{2}\int_{-a}^{a} \sqrt{1+h'^2}\ p_x dx - \int_{-a}^{x} \sqrt{1+h'^2}\ p_x dx \tag{5.12*}$$

A corresponding procedure for equation (4.16) with the conditions (5.9b) yields the solution for s_1 and the constraints to be satisfied by the edge quantities $s_1^{\mp}$ and

$d_1^{\mp}$. Rather than perform the double integration of equation (4.16), it is more convenient to integrate successively the equivalent system (4.14) and (4.13) and apply the conditions (5.3b) and (5.3c). Integrating equation (4.14) over the beam length, we note the edge conditions (5.3c): setting

$$f_0^*(x) = -\int_{-a}^{x} [\sqrt{1+h'^2}\; p + 2hf_0]\, dx \tag{5.13}$$

we obtain the constraints on $\tau_0^{\mp}$ in the form

$$2(h^+ \overset{+}{\tau}_0 - h^- \overset{-}{\tau}_0) = -f_0^*(a) \tag{5.14}$$

The indefinite integral of (4.14) together with (5.14) then gives the alternative forms for the coefficient τ_0,

$$2h\tau_0 = 2h^- \overset{-}{\tau}_0 - f_0^*(x) \tag{5.15}$$

$$2h\tau_0 = h^+ \overset{+}{\tau}_0 + h^- \overset{-}{\tau}_0 + \frac{1}{2} f_0^*(a) - f_0^*(x) \tag{5.15*}$$

We may now deal in a similar manner with equation (4.13). If we substitute for τ_0 from (5.15*) into (4.13) and, after integrating over the beamlength, we apply the edge conditions (5.3b), we obtain the constraining relation

$$\frac{2}{3}[(h^+)^2 \overset{+}{s}_1 - (h^-)^2 \overset{-}{s}_1] = 2a[h^+ \overset{+}{\tau}_0 + h^- \overset{-}{\tau}_0 + \frac{1}{2} f_0^*(a)] - \int_{-a}^{a} [h\sqrt{1+h'^2}\; p_x^* + f_0^*(x)]\, dx \tag{5.16}$$

on the coefficients $s_1^{\mp}$. The corresponding indefinite integration together with the application of the first of conditions (5.3b) yields the solution

$$\frac{2}{3}h^2 s_1 = \frac{2}{3}(h^-)^2 \overset{-}{s}_1 + [h^+ \overset{+}{\tau}_0 + h^- \overset{-}{\tau}_0 + \frac{1}{2} f_0^*(a)](x+a) - \int_{-a}^{x} [h\sqrt{1+h'^2}\; p_x^* + f_0^*(x)]\, dx \tag{5.17}$$

for the coefficient s_1: this latter formula could also be put in a more symmetric form by utilizing relation (5.16).

Relations (5.12), (5.15) and (5.17) are the complete solutions for the coefficients s_0, τ_0 and s_1, respectively: these are the principal stresses considered in classical beam theory. The conditions (5.11), (5.14) and (5.16) on the coefficients appearing in the edge distributions ensure overall static equilibrium.

6. Representation for the Displacement Components

As the Legendre series for the stresses were derived from the equilibrium equations (2.9), so shall the corresponding representations for the displacements be obtained from the second and third of the constitutive relations (2.10). We shall confine our analysis to homogeneous media and shall assume that the body force density has no axial variation. This restriction, to the case where the elastic parameters E and ν are constant, results in a considerable moderation in the algebraic manipulation, while the minor assumption that f_0 is constant permits some simplification in the notation.

If the expansion (3.4) for σ_x and the representation (3.12) for σ are introduced into the third constitutive relation (2.10c), there results the Legendre series for $w,_t$, namely

$$w,_t = \frac{h}{E} \sum_{n=0}^{\infty} (\sigma_n - \nu s_n) P_n - \frac{h^2}{E} f_0 P_1 \tag{6.1}$$

which, with the aid of (3.2), can be immediately integrated to give

$$w = w_0 - \frac{h}{E} \sum_{n=1}^{\infty} \left[\left(\frac{\sigma_{n+1}}{2n+3} - \frac{\sigma_{n-1}}{2n-1} \right) - \left(\frac{s_{n+1}}{2n+3} - \frac{s_{n-1}}{2n-1} \right) \right] P_n - \frac{h^2}{3E} f_0 P_2 \tag{6.2}$$

where w_0 is an unknown function of x introduced by the integration. With the notation

$$\left.\begin{aligned} w_2 &= -\frac{h}{E}\left[\left(\frac{\sigma_3}{7} - \frac{\sigma_1}{3} \right) - \nu \left(\frac{s_3}{7} - \frac{s_1}{3} \right) \right] - \frac{h^2}{3E} f_0 \\ w_n &= -\frac{h}{E}\left[\left(\frac{\sigma_{n+1}}{2n+3} - \frac{\sigma_{n-1}}{2n-1} \right) - \nu \left(\frac{s_{n+3}}{2n+3} - \frac{n-1}{2n-1} \right) \right], \\ & \qquad n \geq 1,\ n \neq 2 \end{aligned}\right\} \tag{6.3}$$

the expansion (6.2) takes the form

$$w = \sum_{n=0}^{\infty} w_n(x)\, P_n(t) \tag{6.4}$$

With a view to obtaining expansions for ω and u, we next compute the derivatives of w appearing in (2.11) and in the second constitutive relation (2.10b). From (6.2) we have that

$$w,_x = w_0' - \frac{1}{E} \sum_{n=1}^{\infty} \left[\left(\frac{h\sigma_{n+1}}{2n+3} - \frac{h\sigma_{n-1}}{2n-1} \right)' - \nu \left(\frac{hs_{n+1}}{2n+3} - \frac{hs_{n-1}}{2n-1} \right)' \right] P_n - \frac{2hh'}{3E} f_0 P_2 \tag{6.5}$$

Referring to relation (3.10*), it can be easily checked that

$$\left(\frac{hs_{n+1}}{2n+3} - \frac{hs_{n-1}}{2n-1}\right)' = \tau_n - h'\left(\frac{n+1}{2n+3}s_{n+1} + \frac{n}{2n-1}s_{n-1}\right), \; n \geq 1 \quad \textbf{(6.6)}$$

which, when inserted into (6.5), gives

$$w,_x = w_0' - \frac{2hh'}{3E} f_0 P_2$$

$$-\frac{1}{E}\sum_{n=1}^{\infty}\left[\left(\frac{h\sigma_{n+1}}{2n+3} - \frac{h\sigma_{n-1}}{2n-1}\right)' + \nu h'\left(\frac{n+1}{2n+3}s_{n+1} + \frac{n}{2n-1}s_{n-1}\right) - \nu\tau_n\right]P_n \quad \textbf{(6.7)}$$

Next from (6.1) we note that

$$tw,_t = \frac{h}{E}\sum_{n=0}^{\infty}(\sigma_n - \nu s_n)\, tP_n - \frac{h^2}{E} f_0 tP_1 \quad \textbf{(6.8)}$$

in which we now use the recurrence relation*

$$tP_n = \frac{n+1}{2n+1}P_{n+1} + \frac{n}{2n+1}P_{n-1} \quad \textbf{(6.9)}$$

to express the right hand side as a Legendre series: a rearrangement then gives

$$\frac{h'}{h}tw,_t = \frac{1}{E}\sum_{n=0}^{\infty}\left[h'\left(\frac{n+1}{2n+3}\sigma_{n+1} + \frac{n}{2n-1}\sigma_{n-1}\right)\right.$$

$$\left.-\nu h'\left(\frac{n+1}{2n+3}s_{n+1} + \frac{n}{2n-1}s_{n-1}\right)\right]P_n - \frac{hh'}{3E}f_0(2P_2 + P_0) \quad \textbf{(6.10)}$$

Combining relations (6.7) and (6.10) and noting the cancellations, we rearrange and find

$$\frac{h'}{h}tw,_t - w,_x = \left[\frac{h'}{3E}(\sigma_1 - \nu s_1 - hf_0) - w_0'\right]$$

$$+\frac{1}{E}\sum_{n=1}^{\infty}\left[h\left(\frac{\sigma_{n+1}'}{2n+3} - \frac{\sigma_{n-1}'}{2n-1}\right) + h'\left(\frac{n+2}{2n+3}\sigma_{n+1} + \frac{n-1}{2n-1}\sigma_{n-1}\right) - \nu\tau_n\right]P_n \quad \textbf{(6.11)}$$

* See footnote on p.16

If we introduce the expansion (6.11), together with the representation (3.9) for τ, into the right hand side of (2.11), we obtain

$$\omega = \frac{1}{E}\left[(1+\nu)\,\tau_0 + \frac{h'}{3}(\sigma_1 - s_1 - hf_0)\right] - w_0'$$
$$+ \frac{1}{E}\sum_{n=1}^{\infty}\left[\tau_n + h\left(\frac{\sigma'_{n+1}}{2n+3} - \frac{\sigma'_{n-1}}{2n-1}\right) + h'\left(\frac{n+2}{2n+3}\,\sigma_{n+1} + \frac{n-1}{2n-1}\,\sigma_{n-1}\right)\right] P_n \tag{6.12}$$

which may be written in the form

$$\omega = \sum_{n=0}^{\infty} \omega_n(x)\, P_n(t) \tag{6.13}$$

with the notation

$$\omega_0 = \frac{1}{E}\left[(1+\nu)\,\tau_0 + \frac{h'}{3}(\sigma_1 - s_1 - hf_0)\right] - w_0' \tag{6.14a}$$

$$\omega_n - \frac{1}{E}\left[\tau_n + h\left(\frac{\sigma'_{n+1}}{2n+3} - \frac{\sigma'_{n-1}}{2n-1}\right) + h'\left(\frac{n+2}{2n+3}\,\sigma_{n+1} + \frac{n-1}{2n-1}\,\sigma_{n-1}\right)\right], \quad n \geq 1. \tag{6.14b}$$

If we introduce τ_0 from (4.17) and τ_n, $n \geq 1$ from (3.10*) into (6.14), we obtain the alternate form

$$\omega_0 = \frac{1}{3E}\left\{(1+\nu)\left[hs_1' + \frac{3}{2}\sqrt{1 + h'^2}\; p^*\right] + h'\left[(2+\nu)s_1 + \sigma_1 - hf_0\right]\right\} - w_0' \tag{6.14a*}$$

$$\omega_n = \frac{1}{E}\left\{h\left[\frac{s'_{n+1} + \sigma'_{n+1}}{2n+3} - \frac{s'_{n-1} + \sigma'_{n-1}}{2n-1}\right] + h'\left[\frac{n+2}{2n+3}(s_{n+1} + \sigma_{n+1}) + \frac{n-1}{2n-1}(s_{n-1} + \sigma_{n-1})\right]\right\}, \quad n \geq 1. \tag{6.14b*}$$

giving the ω_n directly in terms of the s_n and σ_n.

In a similar manner, if we insert the expansions (6.11) and (3.9) into the second constitutive relation (2.10b), we find

$$u_{,t} = \left[\frac{h}{E} [2(1+\nu)\,\tau_0 + \frac{h'}{3}(\sigma_1 - \nu s_1 - h f_0)] - h w_0' \right]$$

$$+ \frac{h}{E} \sum_{n=1}^{\infty} \left[(2+\nu)\,\tau_n + h \left(\frac{\sigma_{n+1}'}{2n+3} - \frac{\sigma_{n-1}'}{2n-1} \right) + h' \left(\frac{n+2}{2n+3}\sigma_{n+1} + \frac{n-1}{2n-1}\sigma_{n-1} \right) \right] P_n \tag{6.15}$$

which integrates to give the Legendre series for u, namely

$$u = \sum_{n=0}^{\infty} u_n(x)\, P_n(t) \tag{6.16}$$

where u_0 is an unknown function of x introduced by the integration and the higher coefficients are given as follows

$$u_1 = -h w_0' + \frac{h}{E} \left\{ 2(1+\nu)\,\tau_0 - \frac{1}{5}(2+\nu)\,\tau_2 - \frac{h}{5}\left(\frac{\sigma_3'}{7} - \frac{\sigma_1'}{3}\right) \right.$$

$$\left. - h' [\frac{4}{5}(\frac{\sigma_3}{7} - \frac{\sigma_1}{3}) + \frac{1}{3}(\nu s_1 + h f_0)] \right\} \tag{6.17a}$$

$$u_n = \frac{h}{E} \left\{ (2+\nu) \left[\frac{\tau_{n-1}}{2n-1} - \frac{\tau_{n+1}}{2n+3} \right] \right.$$

$$- h \left[\frac{\sigma_{n-2}'}{(2n-3)(2n-1)} - \frac{2\sigma_n'}{(2n-1)(2n+3)} + \frac{\sigma_{n+2}'}{(2n+3)(2n+5)} \right]$$

$$\left. + h' \left[\frac{(n-2)\sigma_{n-2}}{(2n-3)(2n-1)} + \frac{3\sigma_n}{(2n-1)(2n+3)} - \frac{(n+3)\sigma_{n+2}}{(2n+3)(2n+5)} \right] \right\},\ n \geq 2\,. \tag{6.17b}$$

If, in the above formulae, we introduce τ_0 from (4.17) and the remaining τ_n $(n \geq 1)$ from (3.10*), we obtain expressions for the u_n directly in terms of the s_n and σ_n, namely

$$u_1 = -h w_0' + \frac{h}{E} \left\{ h \left[\frac{(12+11\nu)s_1' + \sigma_1'}{15} - \frac{(2+\nu)s_3' + \sigma_3'}{35} \right] + \frac{1}{2}\sqrt{1 + h'^2}\, p_x^* \right.$$

$$\left. + h' \left[\frac{2(9+7\nu)s_1 + 4\sigma_1}{15} - \frac{4(2+\nu)s_3 + 4\sigma_3}{35} - \frac{h f_0}{3} \right] \right\} \tag{6.17a\,*}$$

and for $n \geq 2$,

$$u_n = -\frac{h}{E}\left\{ h\left[\frac{(2+\nu)s'_{n-2}+\sigma'_{n-2}}{(2n-3)(2n-1)} - 2\frac{(2+\nu)s'_n+\sigma'_n}{(2n-1)(2n+3)} + \frac{(2+\nu)s'_{n+2}+\sigma'_{n+2}}{(2n+3)(2n+5)}\right]\right.$$

$$\left. -h'\left[\frac{(n-2)[(2+\nu)s'_{n-2}+\sigma'_{n-2}]}{(2n-3)(2n-1)} + \frac{3[(2+\nu)s_n+\sigma'_n]}{(2n-1)(2n+3)} - \frac{(n+3)[(2+\nu)s_{n+2}+\sigma_{n+2}]}{(2n+3)(2n+5)}\right]\right\},$$

$$n \geq 2. \qquad \textbf{(6.17b*)}$$

The expansions (6.4) and (6.16) with the coefficients given respectively by (6.3) and (6.17) are the Legendre series for the displacements: the corresponding representation for the rotation is given by (6.13) with the coefficients determined by relations (6.14). We note that, except for the two undetermined functions u_0 and w_0, all the coefficients in the above Legendre series have been determined in terms of the coefficients appearing in the series for the stresses. From the relations

$$\frac{1}{2h}\int_{-h}^{h} v_x dz = \frac{1}{2}\int_{-1}^{1} u dt = u_0, \quad \frac{1}{2h}\int_{-h}^{h} v_z dz = \frac{1}{2}\int_{-1}^{1} w dt = w_0 \qquad \textbf{(6.18)}$$

we see that u_0 and w_1 measure the average through the thickness of the axial and transverse displacement components respectively and so we may refer to them as the mean displacements: similarly ω_0 measures the mean rotation.

Hence, we have that the displacement quantities are completely determined in terms of the mean displacements and the coefficients in the representations for the stresses. Once the stresses have been determined, the completion of the solution lies in the determination of the mean displacements; in fact, the latter are the only displacement quantities that appear explicitly in the analysis.

7. The Equations for the Unknown Function

The equations to be satisfied by the undetermined functions (namely u_0, w_0 and s_n, $n \geq 2$) will follow from the introduction of the Legendre expansions for the stresses and displacements into the first constitutive relation (2.10a). The orthogonality of the Legendre polynomials then requires that this equation be satisfied term by term, leading to an infinite sequence of equations to be satisfied by the coefficients. The first two are the equations for the determination of the mean displacements u_0 and w_0, respectively, while the remainder constitute the set of equations to be satisfied by the sequence of coefficients $\{s_n,\ n \geq 2\}$.

We start with the calculation of the factors on the left of equation (2.10a). For the first factor we have from (6.16) that

$$u_{,x} = \sum_{n=0}^{\infty} u_n'(x)\, P_n(t) \tag{7.1}$$

in which we may substitute for the u_n' from the formulae resulting from the differentiation of relations (6.17): after some rearrangement, we find that

$$\begin{aligned}
hu_{,x} &= hu_0'P_0 \\
&+ \frac{h}{E}\Bigg\{ \nu(h\tau_0)' + (2+\nu)\left(h\tau_0 - \frac{2}{5}h\tau_2\right)' + \frac{h^2}{5}\left(\frac{\sigma_1'}{3} - \frac{\sigma_3'}{7}\right) + hh'\left[\frac{6}{5}\left(\frac{\sigma_1'}{3} + \frac{\sigma_3'}{7}\right) - \nu\frac{s_1'}{3}\right] \\
&\quad + (hh'' + h'^2)\left[\frac{4}{5}\left(\frac{\sigma_1}{3} - \frac{\sigma_3}{7}\right) - \nu\frac{s_1}{3}\right] - \frac{1}{3}(hh'' + 2h'^2)hf_0 - E(hw_0'' + h'w_0')\Bigg\} P_1 \\
&+ \frac{h}{E}\sum_{n=2}^{\infty}\Bigg\{ (2+\nu)\left[\frac{h\tau_{n-1}}{2n-1} - \frac{h\tau_{n+1}}{2n+3}\right] \\
&\quad - h^2\left[\frac{\sigma_{n-2}''}{(2n-3)(2n-1)} - \frac{2\sigma_n''}{(2n-1)(2n+3)} + \frac{\sigma_{n+2}''}{(2n+3)(2n+5)}\right] \\
&\quad + hh'\left[\frac{(n-4)\sigma_{n-2}'}{(2n-3)(2n-1)} + \frac{7\sigma_n'}{(2n-1)(2n+3)} - \frac{(n+5)\sigma_{n+2}''}{(2n+3)(2n+5)}\right] \\
&\quad + (hh'' + h'^2)\left[\frac{(n-2)\sigma_{n-2}}{(2n-3)(2n-1)} + \frac{3\sigma_n}{(2n-1)(2n+3)} - \frac{(n+3)\sigma_{n+2}}{(2n+3)(2n+5)}\right]\Bigg\} P_n
\end{aligned} \tag{7.2}$$

It can be readily checked from (3.13*) that

$$\left[\frac{h\tau_{n-1}}{2n-1} - \frac{h\tau_{n+1}}{2n+3}\right]' = h'\left[\frac{n}{2n-1}\tau_{n-1} + \frac{n+1}{2n+3}\tau_{n+1}\right] - \sigma_n, \; n \geq 1 \tag{7.3}$$

which, when introduced into (7.2), gives

$$
\begin{aligned}
hu,_x &= hu_0' P_0 \\
&+ \frac{h}{E}\Bigg\{ \nu(h\tau_0)' + (2+\nu)h'\left(\tau_0 + \frac{2}{5}\tau_2\right) - (2+\nu)\sigma_1 + \frac{h^2}{5}\left(\frac{\sigma_1''}{3} - \frac{\sigma_3''}{7}\right) \\
&\quad + hh'\left[\frac{6}{5}\left(\frac{\sigma_1'}{3} - \frac{\sigma_3'}{7}\right) - \nu\frac{s_1'}{3}\right] \\
&\quad + (hh'' + h'^2)\left[\frac{4}{5}\left(\frac{\sigma_1}{3} - \frac{\sigma_3}{7}\right) - \nu\frac{s_1}{3}\right] - \frac{1}{3}(hh'' + 2h'^2)hf_0 - E(hw_0'' + h'w_0')\Bigg\} P_1 \\
&+ \frac{h}{E}\sum_{n=2}^{\infty}\Bigg\{ (2+\nu)h'\left[\frac{n}{2n-1}\tau_{n-1} + \frac{n+1}{2n+3}\tau_{n+1}\right] - (2+\nu)\sigma_n \\
&\quad - h^2\left[\frac{\sigma_{n-2}''}{(2n-3)(2n-1)} - \frac{2\sigma_n''}{(2n-1)(2n+3)} + \frac{\sigma_{n+2}''}{(2n+3)(2n+5)}\right] \\
&\quad + hh'\left[\frac{(n-4)\sigma_{n-2}'}{(2n-3)(2n-1)} - \frac{7\sigma_n'}{(2n-1)(2n+3)} + \frac{(n+5)\sigma_{n+2}'}{(2n+3)(2n+5)}\right] \\
&\quad + (hh'' + h'^2)\left[\frac{(n-2)\sigma_{n-2}}{(2n-3)(2n-1)} - \frac{3\sigma_n}{(2n-1)(2n+3)} + \frac{(n+3)\sigma_{n+2}}{(2n+3)(2n+5)}\right]\Bigg\} P_n
\end{aligned}
\tag{7.4}
$$

For the calculation of the second factor on the left of (2.10a), we observe that, when multiplied by t, equation (6.15) gives

$$
\begin{aligned}
tu,_t &= \frac{h}{E}[2(1+\nu)\tau_0 + \frac{1}{3}h'(\sigma_1 - \nu s_1 - hf_0) - Ew_0']P_1 \\
&+ \frac{h}{E}\sum_{n=2}^{\infty}\left[(2+\nu)\tau_n - h\left(\frac{\sigma_{n-1}'}{2n-1} - \frac{\sigma_{n+1}'}{2n+3}\right) + h'\left(\frac{n-1}{2n-1}\sigma_{n-1} + \frac{n+2}{2n+3}\sigma_{n+1}\right)\right] tP_n
\end{aligned}
\tag{7.5}
$$

Applying the recurrence relation (6.9), we rearrange and find

$$h' tu,_t = -\frac{hh'}{3E}\left[\left\{(2+\nu)(hs_0' - \frac{1}{5}hs_2' - \frac{3}{5}h' s_2) + (h\sigma_0' - \frac{1}{5}h\sigma_2' - \frac{3}{5}h'\sigma_2)\right\}P_0\right.$$

$$+\frac{h}{E}\left[\left\{\nu h'\tau_0 + (2+\nu)h'(\tau_0 + \frac{1}{5}\tau_2) - \frac{2}{5}hh'\left(\frac{\sigma_1'}{3} - \frac{\sigma_3'}{7}\right)\right.\right.$$

$$\left.\left. + h'^2\left(\frac{7\sigma_1}{15} + \frac{8\sigma_3}{35} - \frac{\nu s_1}{3}\right) - \frac{1}{3}hh'^2 f_0 - Eh' w_0'\right\}P_1\right]$$

$$+\frac{h}{E}\sum_{n=2}^{\infty}\left[\left\{(2+\nu)h'\left[\frac{n}{2n-1}\tau_{n-1} + \frac{n+1}{2n+3}\tau_{n+1}\right]\right.\right.$$

$$-hh'\left[\frac{n\sigma_{n-2}'}{(2n-3)(2n-1)} - \frac{\sigma_n'}{(2n-1)(2n+3)} + \frac{(n+1)\sigma_{n+2}'}{(2n+3)(2n+5)}\right]$$

$$\left.\left. + h'^2\left[\frac{n(n-2)\sigma_{n-2}}{(2n-3)(2n-1)} + \frac{2n(n+1)\sigma_n}{(2n-1)(2n+3)} - \frac{(n+1)(n+3)\sigma_{n+2}}{(2n+3)(2n+5)}\right]\right\}P_n\right] \tag{7.6}$$

where we have introduced the formula for τ_1 from (4.19) into the coefficient of P_0.

The subtraction of equation (7.6) from equation (7.4) will yield the Legendre series for the left hand side of equation (2.10a): after the subtraction, we introduce formula (4.17) for τ_0 into the coefficient of P_1 and make some rearrangements. If we set

$$\phi_0 = -\frac{h'}{3}[(2+\nu)(hs_0' - \frac{1}{5}hs_2' - \frac{3}{5}h' s_2) + (h\sigma_0' - \frac{1}{5}h\sigma_2' - \frac{3}{5}h'\sigma_2)] \tag{7.7a}$$

$$\phi_1 = -\frac{\nu}{3}[h^2 s_1'' + 2hh' s_1' + hh'' s_1 + \frac{3h}{2}(\sqrt{1+h'^2}\, p_x^*)'] + \frac{h}{3}(hh'' + h'^2)f_0$$

$$-\frac{h^2}{5}\left(\frac{\sigma_1''}{3} - \frac{\sigma_3''}{7}\right) - \frac{8hh'}{5}\left(\frac{\sigma_1'}{3} - \frac{\sigma_3'}{7}\right) + \frac{h'^2}{5}\left(\sigma_1 + \frac{12}{7}\sigma_3\right) - \frac{4hh''}{5}\left(\frac{\sigma_1}{3} - \frac{\sigma_3}{7}\right) \tag{7.7b}$$

$$\phi_n = h^2\left[\frac{\sigma_{n-2}''}{(2n-3)(2n-1)} - \frac{2\sigma_n''}{(2n-1)(2n+3)} + \frac{\sigma_{n+2}''}{(2n+3)(2n+5)}\right]$$

$$-hh'\left[\frac{2(n-s)\sigma'_{n-2}}{(2n-3)(2n-1)}+\frac{6\sigma'_n}{(2n-1)(2n+3)}-\frac{2(n+3)\sigma'_{n+2}}{(2n+3)(2n+5)}\right]$$

$$+h'^2\left[\frac{(n-2)(n-1)\sigma_{n-2}}{(2n-3)(2n-1)}+\frac{(2n^2+2n-3)\sigma_n}{(2n-1)(2n+3)}+\frac{(n+2)(n+3)\sigma_{n+2}}{(2n+3)(2n+5)}\right]$$

$$-hh''\left[\frac{(n-2)\sigma_{n-2}}{(2n-3)(2n-1)}+\frac{3\sigma_n}{(2n-1)(2n+3)}-\frac{(n+3)\sigma_{n+2}}{(2n+3)(2n+5)}\right], n\geq 2. \tag{7.7c}$$

then the Legendre series for the left hand side of (2.10a) is given by

$$hu,_x-h'tu,_t=\frac{h}{E}\{(Eu_0'-\phi_0)P_0-[Ehw_0''+(2+\nu)\sigma_1+\phi_1]P_1-\sum_{n=2}^{\infty}[(2+\nu)\sigma_n+\phi_n]P_n\} \tag{7.8}$$

We note the formal similarity between relations (7.7c) expressing the ϕ_n in terms of the σ_n and relation (3.14) giving the σ_n in terms of the s_n.

We can now write equation (2.10a) in series form by introducing the expansions (3.4) and (3.12) into the right hand side and the series (7.8) into the left; if we make some rearrangements and ignore the common multiplying factor h/E, we find

$$[Eu_0'-(s_0-\nu\sigma_0+\phi_0)]P_0-[Ehw_0''+s_1+2\sigma_1+\phi_1+\nu hf_0]P_1$$
$$-\sum_{n=2}^{\infty}[s_n+2\sigma_n+\phi_n]P_n=0 \tag{7.9}$$

and we note that the terms with Poisson's ratio cancel in all coefficients beyond the first i.e. for $n\geq 2$. The orthogonality of Legendre polynomials now requires that each coefficient on the left of (7.9) vanish independently so that we have

$$Eu_0'=s_0-\nu\sigma_0+\phi_0 \tag{7.10a}$$

$$-Ehw_0''=s_1+2\sigma_1+\phi_1+\nu hf_0 \tag{7.10b}$$

$$0=s_n+2\sigma_n+\phi_n,\ n\geq 2 \tag{7.10c}$$

As will be made clear in the next Section, the system of equation (7.10c) is self-contained and is to be solved subject to the boundary conditions (5.9c) to give the unique solution for the unknown coefficients $\{s_n, n\geq 2\}$. The unknown mean displacements u_0 and w_0 are to be determined from the integration of equations (7.10a) and (7.10b) respectively: for uniqueness in the solution of this latter pair, it is necessary to state a set of conditions on the displacements.

Generally, such conditions follow from the physical requirement that a point together with an element through the point be fixed in the body. Taking this point at the origin and choosing the transverse element through the origin as the fixed element would require that

$$u(0,0) = 0, \quad w(0,0) = 0, \quad \omega(0,0) = 0 \tag{7.11a,b,c}$$

The conditions (7.11) could be replaced by the conditions on the coefficients

$$\sum_{k=0}^{\infty} u_{2k}(0)P_{2k}(0) = 0, \quad \sum_{k=0}^{\infty} w_{2k}(0)P_{2k}(0) = 0, \quad \sum_{k=0}^{\infty} \omega_{2k}(0)P_{2k}(0) = 0 \tag{7.11a,b,c*}$$

since, for the Legendre polynomials of odd order, we have

$$P_{2k+1}(0) = 0 \tag{7.12}$$

If we also note the values at the origin

$$P_{2k}(0) = (-1)^k \frac{(2k-1)!!}{(2k)!!} \tag{7.13}$$

for the polynomials of even order, we may write conditions (7.11*) in the form

$$u_0(0) = -\sum_{k=1}^{\infty} (-1)^k \frac{(2k-1)!!}{(2k)!!} u_{2k}(0) \tag{7.14}$$

$$w_0(0) = -\sum_{k=1}^{\infty} (-1)^k \frac{(2k-1)!!}{(2k)!!} w_{2k}(0) \tag{7.15a}$$

$$\omega_0(0) = -\sum_{k=1}^{\infty} (-1)^k \frac{(2k-1)!!}{(2k)!!} \omega_{2k}(0) \tag{7.15b}$$

If we recall formula (6.14a*) for ω_0, we see that condition (7.15b) may be replaced by a second condition on w_0, namely

$$w_0'(0) = \frac{1}{3E}\{(1+\nu)[h(0)s_1'(0) + \frac{3}{2}\sqrt{1 + h'^2(0)}\, p_x^*(0)]$$

$$+ h'(0)[(2+\nu)s_1(0) + \sigma_1(0) - h(0)f_0] + \sum_{k=1}^{\infty} (-1)^k \frac{(2k-1)!!}{(2k)!!} \omega_{2k}(0) \tag{7.15b*}$$

which may prove more convenient in later use.

8. The Uncoupling of Effects: Principal and Residual Parts

An inspection of equations (7.10) together with relations (3.14a), (4.20) and (7.7) shows that the general problem uncouples into the two distinct problems of stretching and bending which may be characterized as follows:

1. *The Problem of Stretching* is concerned with the quantities with even index in the sequences $\{s_n\}$, $\{\sigma_n\}$ and $\{\phi_n\}$ together with the mean axial displacement u_0: also involved are the coefficients with even index in the sequence $\{u_n\}$ and the quantities with odd index in the sequences $\{\tau_n\}$, $\{w_n\}$ and $\{\omega_n\}$.
2. *The Problem of Bending* is concerned with the quantities with odd index in the sequences $\{s_n\}$, $\{\sigma_n\}$ and $\{\phi_n\}$ together with the mean transverse displacement w_0 and the mean rotation ω_0: also involved are the coefficients with even index in the sequences $\{\tau_n\}$, $\{w_n\}$, $\{\omega_n\}$ together with the quantities with odd index in the sequence $\{u_n\}$:

In particular, the pair of equations (7.10a,b), with their respective boundary conditions (5.9a,b) do not interact.

It has been remarked in Section Seven that there is a further partial uncoupling in the system (7.10): namely, the system (7.10c) is independent of both equations (7.10a,b). This is evident from the following observations:

1. the set of quantities $\{\phi_n, n \geq 2\}$ are expressed entirely in terms of the $\{\sigma_n, n \geq 0\}$; this is evident in the definitions (7.7c):
2. the latter set $\{\sigma_n, n \geq 0\}$ are expressed in terms of the original coefficients $\{s_n, n \geq 0\}$; this has been shown in relations (3.14) and (4.20):
3. the quantities s_0 and s_1 have already been determined from relations (4.12) and (4.16) respectively leading to formulae (5.12) and (5.17).

Hence, the system (7.10c) constitutes a self-contained system of equations subject to the boundary conditions (5.9c), for the determination of the set of coefficients $\{s_n, n \geq 2\}$.

The coefficients s_0 and s_1 measuring respectively the axial stress resultant and stress couple (c.f. relations (4.23a,b)) are the principal axial stress effects. The coefficients $\{s_n, n \geq 2\}$ have no direct influence on either the axial stress resultant or stress couple: for this reason they shall be called the residual axial stress effects. We have seen in Section 5 that the principal axial stress effects can be determined without reference to the constitutive relations. A direct consequence of this is the self-contained independence of the system (7.10c) which, in turn, means that the residual axial stresses can be determined without reference to the equations for the mean displacements.

There remains a one-way dependence in our derived system of equations (7.10); namely, the present form of the equations for the mean displacements (7.10a,b) involve the residual stress effects as well as the principal stress effects. We shall see presently how this partial coupling may be resolved so that each of the problems of stretching and bending can in turn be separated into two independent problems – one purely for the principal effects, the other involving only the residual effects. This will be accomplished by making an appropriate decomposition of the mean displacements into their principal and residual parts. A necessary preliminary to such a decomposition is the clarification of the manner in which the remaining two stress components – namely, the transverse shear and normal stresses – are to be resolved into principal and residual effects. This latter resolution is most conveniently preformed in the context of the linear vector space induced by our expansion procedure. As all three stress quantities fit this pattern, it is in the interest of consistency of procedure to start by restating the decomposition of the axial stress in this context.

The Legendre series expansion induces on the stress quantities a resolution into their component representations in an infinite dimensional linear vector space. The sequence of Legendre polynomials constitute an orthogonal basis spanning the space and each component in the resolution is defined by the corresponding coefficient in the expansion. Using the notation of vector spaces, we may write

$$\sigma_x = \{s_n : 0 \le n \le \infty\} = \{s_0, s_1, s_2, \dots s_n, \dots\} \tag{8.1a}$$

$$\tau = \{\tau_n : 0 \le n \le \infty\} = \{\tau_0, \tau_1, \tau_2, \dots \tau_n, \dots\} \tag{8.1b}$$

$$\begin{aligned} \sigma &= \{\sigma_0, (\sigma_1 - hf_0), [\sigma_n : 2 \le n \le \infty]\} \\ &= \{\sigma_0, (\sigma_1 - hf_0), \sigma_2, \sigma_3, \dots \sigma_n, \dots\} \end{aligned} \tag{8.1c}$$

In this notation the resolution of σ_x into its principal and residual parts takes the form

$$\sigma_x = \overset{P}{\sigma}_x + \overset{R}{\sigma}_x \tag{8.2}$$

where

$$\overset{P}{\sigma}_x = \{s_0, s_1, 0, 0, \dots 0, \dots\} \tag{8.3a}$$

$$\overset{R}{\sigma}_x = \{0, 0, s_2, s_3, \dots s_x, \dots\} \tag{8.3b}$$

In the vector $\overset{P}{\sigma}_x$ representing the principal part, all components of index greater than 1 are zero, while the zero-th and first components coincide with the zero-th and first components of σ_x, respectively. The vector $\overset{R}{\sigma}_x$ representing the residual part complements $\overset{P}{\sigma}_x$ so that zero-ith and first components are zero while for $n \ge 2$ the nth component is s_n.

In making a corresponding decomposition for the other quantities, it is necessary that their principal and residual effects be defined unambiguously. If we take the definitions,

1. any effect induced, exclusively, by the residual axial stress component $\overset{R}{\sigma}_x$, is a *Residual Effect*,
2. all other effects are to be considered *Principal Effects*;

then we have that all residual effects must be expressed exclusively in terms of the set of coefficients $\{s_n : n \ge 2\}$.

In applying the above definitions to the transverse shear stress $\{\tau_n, 0 \le n \le \infty\}$, we first note from (3.10*) that for $3 \le n \le \infty$ the τ_n are purely residual effects: moreover, from (4.17), τ_0 is a principal effect. Hence to perform the resolution, it suffices to decompose the mixed components τ_1 and τ_2 given explicitly by relations (4.19). We set

$$\overset{P}{\tau_1} = -hs_0', \qquad \overset{R}{\tau_1} = \frac{1}{5}(hs_2' + 3h's_2) \tag{8.4a,b}$$

$$\overset{P}{\tau_2} = -\frac{1}{3}(hs_1' - h's_1), \; \overset{R}{\tau_2} = \frac{1}{7}(hs_3' + 4h's_3) \tag{8.5a,b}$$

so that

$$\tau_1 = \overset{P}{\tau_1} + \overset{R}{\tau_1}, \; \tau_2 = \overset{P}{\tau_2} + \overset{R}{\tau_2} \tag{8.6a,b}$$

and the resolution of τ takes the form

$$\tau = \overset{P}{\tau} + \overset{R}{\tau}$$

where

$$\overset{P}{\tau} = \{\tau_0, \overset{P}{\tau_1}, \overset{P}{\tau_2}, 0, 0 \ldots 0, \ldots\} \tag{8.7a}$$

$$\overset{R}{\tau} = \{0, \overset{R}{\tau_1}, \overset{R}{\sigma_2}, \tau_3, \tau_4 \ldots \tau_n \ldots\} \tag{8.7b}$$

The quantity $\overset{P}{\tau}$ is the principal part of the transverse shear stress: in its vector space representation all components with index greater than 2 are zero while the zero-th components is τ_0. The residual transverse shear effect $\overset{R}{\tau}$ has vanishing zero-th component in the vector space representation while the components with index greater than 2 coincide with those of τ.

In making a corresponding resolution of the transverse normal stress, we note from (3.14) that for $4 \leq n \leq \infty$ the σ_n are purely residual effects; the remaining components σ_0, σ_1, σ_2, and σ_3 are mixed effects as is evident from the explicit relations (4.20) and (4.22): an inspection of the latter shows how the resolution is to be made. Noting relations (4.20), we write

$$\overset{P}{\sigma_0} = -\frac{1}{3h}(h^3 s_0')' + \frac{1}{2}\sqrt{1 + h'^2}\, p^* \tag{8.8a}$$

$$= -\frac{1}{3h}(h^2 h' s_0)' + \frac{1}{2}\sqrt{1 + h'^2}\, p^* + \frac{1}{6h}[h^2\sqrt{1 + h'^2}\, p^*]' \tag{8.8a*}$$

$$\overset{R}{\sigma_0} = \frac{1}{15h}(h^3 s_2)'' \tag{8.8b}$$

for the decomposition of σ_0 and

$$\overset{P}{\sigma_l} = -\frac{1}{5}[2h^2 s_1'' + 6hh's_1' + (3hh'' - h'^2)\, s_1 + \frac{5h}{2}(\sqrt{1 + h'^2}\, p_x^*)'] \tag{8.9a}$$

$$\overset{R}{\sigma}_1 = \frac{1}{35}[h^2 s_3'' + 8hh' s_3' + 4(hh'' + 3h'^2)s_3] \tag{8.9b}$$

for the decomposition of σ_1: similarly formulae (4.22) requires that we set

$$\overset{P}{\sigma}_2 = \frac{1}{3} h^2 s_0'' \tag{8.10a}$$

$$\overset{R}{\sigma}_2 = -\frac{h^2}{3}[\frac{2}{7}s_2'' - \frac{1}{21}s_4''] - hh'[\frac{2}{7}s_2' - \frac{10}{63}s_4']$$

$$+ h'^2[\frac{3}{7}s_2 + \frac{20}{63}s_4] - hh''[\frac{1}{7}s_2 - \frac{5}{64}s_4] \tag{8.10b}$$

and

$$\overset{P}{\sigma}_3 = \frac{1}{15}[h^2 s_1'' - 2hh' s_1' + (2h'^2 - hh'')s_1] \tag{8.11a}$$

$$\overset{R}{\sigma}_3 = -\frac{h^2}{9}[\frac{2}{5}s_3'' - \frac{1}{11}s_5''] - \frac{hh'}{3}[\frac{2}{5}s_3' - \frac{4}{11}s_5']$$

$$+ \frac{h'^2}{3}[\frac{7}{5}s_3 + \frac{10}{11}s_4] - \frac{hh'}{3}[\frac{1}{5}s_3 - \frac{2}{11}s_5] \tag{8.11b}$$

for the decomposition of σ_2 and σ_3, respectively. The resolution of σ may then be written

$$\sigma = \overset{P}{\sigma} + \overset{R}{\sigma} \tag{8.12}$$

where

$$\overset{P}{\sigma} = \{\overset{P}{\sigma}_0, \overset{P}{\sigma}_1, \overset{P}{\sigma}_2, \overset{P}{\sigma}_3, 0, 0 \ldots 0, \ldots..\} \tag{8.13a}$$

$$\overset{R}{\sigma} = \{\overset{R}{\sigma}_0, \overset{R}{\sigma}_1, \overset{R}{\sigma}_2, \overset{R}{\sigma}_3, \sigma_4, \sigma_5, \ldots \sigma_n \ldots\} \tag{8.13b}$$

The quantity $\overset{P}{\sigma}$ is the principal part of the transverse normal stress: in its vector space representation, the components with index greater than 3 are zero. In the corresponding representation of the residual transverse normal stress $\overset{R}{\sigma}$, the components with index greater than 3 coincide with those of σ.

We could now make a similar decomposition on the quantity $\phi = \{\phi_n, n \geq 0\}$: however, in the sequel we shall have occasion to use the principal and residual parts of merely the zero-th and first components and so we limit ourselves to listing these.

Referring to the formula (7.7a), we set

$$\overset{P}{\phi_0} = -\frac{h'}{3}[(2+\nu)hs_0' + h(\overset{P}{\sigma_0} - \frac{1}{5}\overset{P}{\sigma_2})' - \frac{3}{5}h'\overset{P}{\sigma_2}] \tag{8.14a}$$

$$\overset{R}{\phi_0} = \frac{h'}{3}[\frac{(2+\nu)}{5}(hs_2' + 3h's_2) - h(\overset{R}{\sigma_0} - \frac{1}{5}\overset{R}{\sigma_2})' + \frac{3}{5}h'\overset{R}{\sigma_2}] \tag{8.14b}$$

so that

$$\phi_0 = \overset{P}{\phi_0} + \overset{R}{\phi_0} \tag{8.15}$$

Similarly, guided by formula (7.7b) we write

$$\overset{P}{\phi_1} = -\frac{\nu}{3}[h^2 s_1'' + 2hh's_1' + hh''s_1 + \frac{3h}{2}(\sqrt{1+h'^2}\,\dot{p}_x)'] + \frac{h}{3}(hh'' + h'^2)f_0$$
$$-\frac{h^2}{5}\left(\frac{\overset{P}{\sigma_1}}{3} - \frac{\overset{P}{\sigma_3}}{7}\right)'' - \frac{8hh'}{5}\left(\frac{\overset{P}{\sigma_1}}{3} - \frac{\overset{P}{\sigma_3}}{7}\right)' + \frac{h'^2}{5}\left(\overset{P}{\sigma_1} + \frac{12}{7}\overset{P}{\sigma_3}\right) - \frac{4hh''}{5}\left(\frac{\overset{P}{\sigma_1}}{3} - \frac{\overset{P}{\sigma_3}}{7}\right) \tag{8.16a}$$

$$\overset{R}{\phi_1} = -\frac{h^2}{5}\left(\frac{\overset{R}{\sigma_1}}{3} - \frac{\overset{R}{\sigma_3}}{7}\right)'' - \frac{8hh'}{5}\left(\frac{\overset{R}{\sigma_1}}{3} - \frac{\overset{R}{\sigma_3}}{7}\right)' + \frac{h'^2}{5}\left(\overset{R}{\sigma_1} + \frac{12}{7}\overset{R}{\sigma_3}\right) - \frac{4hh''}{5}\left(\frac{\overset{R}{\sigma_1}}{3} - \frac{\overset{R}{\sigma_3}}{7}\right) \tag{8.16b}$$

so that

$$\phi_1 = \overset{P}{\phi_1} + \overset{R}{\phi_1} \tag{8.17}$$

A similar decomposition can be made in the representations for the displacement components. For the higher coefficients, namely those with suffix $n \geq 1$, the resolution follows from observing the formulae of Section Six. Writing

$$u_n = \overset{P}{u_n} + \overset{R}{u_n}, \quad w_n = \overset{P}{w_n} + \overset{R}{w_n}, \; n \geq 1 \tag{8.18a,b}$$

$$\omega_n = \overset{P}{\omega_n} + \overset{R}{\omega_n}, \; n \geq 1 \tag{8.18c}$$

we note, in particular, that

$$\overset{R}{u_n} = u_n, \; n \geq 7 \tag{8.19a}$$

$$\overset{R}{w_n} = w_n, \quad \overset{R}{\omega_n} = \omega_n, \; n \geq 6 \tag{8.19b,c}$$

and the decomposition of the $\{u_n, 1 \leq n \leq 6\}$, $\{w_n, 1 \leq n \leq 5\}$ and $\{\omega_n, 1 \leq n \leq 5\}$

follow from the inspection of formulae (6.17*), (6.3) and (6.14*), respectively. As we shall not need these resolved formulae explicitly in the sequel, we do not exhibit them here.*

The way is now clear for the resolution of the zero-th coefficients representing the mean displacements and the mean rotation into their principal and residual parts in accordance with the definitions given earlier. Starting with the axial displacement, we set

$$u_0 = U + U_R \tag{8.20}$$

where noting (7.10a) and (7.14) the residual component U_R must satisfy

$$EU'_R = -\nu \overset{R}{\sigma_0} + \overset{R}{\phi_0}, \tag{8.21a}$$

$$U_R(0) = -\sum_{k=1}^{\infty} (-1)^k \frac{(2k-1)!!}{(2k)!!} \overset{R}{u}_{2k}(0) \tag{8.21b}$$

The equation to be satisfied by the complementary principal component U follows from the subtraction of equation (8.21a) from equation (7.10a); the associated initial condition is obtained by subtracting (8.21b) from (7.14): we find

$$EU' = s_0 - \nu \overset{P}{\sigma_0} + \overset{P}{\phi_0} \tag{8.22a}$$

$$U(0) = -\sum_{k=1}^{6} (-1)^k \frac{(2k-1)!!}{(2k)!!} \overset{P}{u}_{2k}(0) \tag{8.22b}$$

where in the, latter, we have noted relations (8.19a). We recall that these initial value problems (8.21) and (8.22) are associated with problems of stretching.

The mean transverse displacement and mean rotation, associated with the problem of bending will be decomposed by referring to equations (7.10b), (6.14a*) and conditions (7.15). Starting with formula (6.14a*) for ω_0, we set

$$\omega_0 = \Omega + \Omega_R \tag{8.23}$$

and the principal and residual components are given respectively by

$$\Omega = \frac{1}{3E} \{(1+\nu)\,[hs'_1 + \tfrac{3}{2}\sqrt{1 + h'^2}\, p_x^*] + h'\,[(2+\nu)s_1 + \overset{P}{\sigma_1} - hf_0]\} - W' \tag{8.24a}$$

* In the corresponding treatment of the displacement boundary value problems, it would be necessary to resolve the boundary conditions into the two sets to be associated, respectively, with the principal and residual problems: this decomposition would follow from an inspection of these formulae for the resolution of the displacement coefficients.

$$\Omega_R = \frac{h'}{3E}\sigma_1^R - W_R' \tag{8.24b}$$

where, by addition, we clearly have (recalling (6.14a*))

$$w_0 = W + W_R \tag{8.25}$$

In the above W and W_R are respectively the principal and residual components of the mean axial displacement: for the determination of these quantities, we have from (7.10b) and (7.15) that W_R must satisfy

$$-EhW_R'' = 2\sigma_1^R + \phi_1^R \tag{8.26a}$$

subject to

$$W_R(0) = -\sum_{k=1}^{\infty}(-1)^k\frac{(2k-1)!!}{(2k)!!}w_{2k}^R(0),\ \Omega_R(0) = -\sum_{k=1}^{\infty}(-1)^k\frac{(2k-1)!!}{(2k)!!}\omega_{2k}^R(0) \tag{8.26b,c}$$

and, from (8.24b), we note the alternate form for the condition on Ω_R,

$$W_R'(0) = \frac{h'(0)}{3E}\sigma_1^R(0) + \sum_{k=1}^{\infty}(-1)^k\frac{(2k-1)!!}{(2k)!!}\omega_{2k}^R(0) \tag{8.26c*}$$

By subtracting relations (8.26) respectively from the corresponding relations (7.10b) and (7.15), we obtain, for the principal components,

$$-EhW'' = s_1 + 2\sigma_1^P + \phi_1^P + \nu h f_0 \tag{8.27a}$$

$$W(0) = -\sum_{k=1}^{5}(-1)^k\frac{(2k-1)!!}{(2k)!!}w_{2k}^P(0),\ \Omega(0) = -\sum_{k=1}^{5}(-1)^k\frac{(2k-1)!!}{(2k)!!}\omega_{2k}^P(0) \tag{8.27b,c}$$

where we have noted relations (8.19b,c): from (8.24a), the latter condition on Ω may be replaced by

$$W'(0) = \frac{1}{3E}\{(1+\nu)[h(0)s_1'(0) + \frac{3}{2}\sqrt{1+h'^2(0)}\,p_x^*(0)]$$

$$+ h'(0)[(2+\nu)s_1(0) + \sigma_1^P(0) - h(0)f_0]\} + \sum_{k=1}^{5}(-1)^k\frac{(2k-1)!!}{(2k)!!}\omega_{2k}^P(0) \tag{8.27c*}$$

The systems (8.26) and (8.27) lead to the unique determination of the principal and residual components of the transverse displacement.

This concludes the analysis in the general case. One feature of the resulting equations is immediately evident, namely, the elastic parameters appear only in the equations for the displacements. Specifically, the constants E and ν appear neither in the formulae for the principal stresses of Section Four nor in the equations for the residual stresses (7.10c); although the latter system was derived from the constitutives relations, the constant E disappeared as a common factor and the terms with ν cancelled out. Hence, we have that the equations for the residual stresses are invariant for all *homogeneous isotropic* materials. Moreover, the solution of the stress boundary value problem is independent of the elastic constants. This latter remark does not apply either to the displacement or to the mixed boundary value problems.

We now proceed to an examination of the problem of bending and stretching individually. In order to emphasize the main points of the procedure, we shall confine our attention to the restricted case of the beam with constant thickness so that

$$h' = 0 \tag{8.28}$$

If we note relations (3.14) and (7.7c), we see that this unessential restriction results in a considerable reduction in the algebra, and thus helps to clarify the essential features of the problem.

9. The Problem of Stretching: Restricted Case

In the problem of stretching, the surface stresses are so distributed that

$$p = 0, \; p_x^* = 0 \qquad \textbf{(9.1a,b)}$$

and, in the absence of a mean transverse body force, we have

$$f_0 = 0 \qquad \textbf{(9.1c)}$$

The specification of the edge stress vectors is such that, in their dependence on the thickness coordinate, the normal components are even, while the shear components are odd: hence

$$s_{2k+1}^{\mp} = 0, \; \tau_{2k}^{\mp} = 0, \; k \geq 0 \qquad \textbf{(9.2)}$$

which, combined with (9.1b) and (5.8b), is equivalent to

$$s_{2k+1}^{\mp} = 0, \; d_{2k+1}^{\mp} = 0, \; k \geq 0 \qquad \textbf{(9.3)}$$

From the above relations, it follows that the coefficients associated with the bending effects satisfy homogeneous differential equations with homogeneous boundary conditions and therefore vanish identically: explicitly, we have

$$s_{2k+1}(x) \equiv 0, \; \sigma_{2k+1}(x) \equiv 0, \; u_{2k+1}(x) \equiv 0, \; k \geq 0 \tag{9.4a}$$

$$\tau_{2k}(x) \equiv 0, \quad w_{2k}(x) \equiv 0, \quad \omega_{2k}(x) \equiv 0, \; k \geq 0 \tag{9.4b}$$

$$\phi_{2k+1}(x) \equiv 0, \; k \geq 0 \tag{9.4c}$$

Accordingly, the expansions for the stress components take the form

$$\sigma_x = \sum_{k=0}^{\infty} s_{2k}(x) P_{2k}(t) \tag{9.5a}$$

$$\tau = \sum_{k=0}^{\infty} \tau_{2k+1}(x) P_{2k+1}(t) \tag{9.5b}$$

$$\sigma = \sum_{k=0}^{\infty} \sigma_{2k}(x) P_{2k}(t) \tag{9.5c}$$

The zero-th components in these expansions satisfy the pair of equations.

$$2hs_0' + p_x = 0 \tag{9.6}$$

$$\frac{2}{3} h\tau_1' - 2\sigma_0 + p^* = 0 \tag{9.7}$$

which follow from the introduction of the restriction

$$h' = 0 \tag{9.8}$$

into equations (4.12) and (4.15), respectively. Relation (9.6) is the equation for the determination of the principal axial stress coefficient s_0: the equation satisfied by the residual axial stress coefficients $\{s_{2k}, k \geq 1\}$ in expansions (9.5a) will be stated later.

The shear stress coefficients in the expansion (9.5b) are given in terms of the axial stress coefficients by the system of relations

$$\tau_{2k+1} = h \left[\frac{s_{2k+2}'}{4k+5} - \frac{s_{2k}'}{4k+1} \right], \; k \geq 0 \tag{9.9}$$

which follows from the introduction of (9.8) into relations (9.10) with $n = 2k + 1$.

In terms of the shear stress coefficients, the transverse normal stress coefficients in the expansions (9.5c) are given by

$$\sigma_0 = \frac{1}{3} h\tau_1' + \frac{1}{2} p^* \tag{9.10a}$$

$$\sigma_{2k} = h\left[\frac{\tau'_{2k+1}}{4k+3} - \frac{\tau'_{2k-1}}{4k-1}\right], \quad k \geq 1 \tag{9.10b}$$

where the latter follows from the introduction of (9.8) into relations (3.13) with $n = 2k$, while the former is a rearrangement of equation (9.7). The introduction of (9.9) into (9.10) gives the corresponding expressions in terms of the axial stress coefficients: relation (9.10a) yields the restricted form ($h' = 0$) of (4.20a) while the set (9.10b) becomes the restricted form of relations (3.14) with even index: written explicitly, we have

$$\sigma_0 = h^2\left[-\frac{1}{3}s_0'' + \frac{1}{15}s_2''\right] + \frac{1}{2}p^* \tag{9.11a}$$

$$\sigma_{2k} = h^2\left[\frac{s''_{2k-2}}{(4k-3)(4k-1)} - \frac{2s''_{2k}}{(4k-1)(4k+3)} + \frac{s''_{2k+2}}{(4k+3)(4k+5)}\right], \quad k \geq 1 \tag{9.11b}$$

We may write the sequence of equations (9.9) explicitly for successive values of k and perform the summation: noting the cancellations and substituting for hs_0' from (9.6), we obtain the analog, for the restricted case, of relations (5.7), namely, the sequence

$$hs'_{2k} = (4k+1)\sum_{j=0}^{k-1}\tau_{2j+1} - \frac{1}{2}p_x, \quad k \geq 1 \tag{9.12}$$

for which relation (9.6) is the representative for $k = 0$. Thereby, the quantities $d_{2k}^{\mp}$, defined by

$$d_{2k}^{\mp} = (4k+1)\sum_{j=0}^{k-1}\tau_{2j+1}^{\mp} - \frac{1}{2}p_x(\mp a), \quad k \geq 1 \tag{9.13}$$

enable us to transform the edge conditions

$$s_{2k}(\mp a) = s_{2k}^{\mp}, \quad \tau_{2k+1}(\mp a) = \tau_{2k+1}^{\mp}, \quad k \geq 0 \tag{9.14}$$

into the equivalent system

$$s_0(\mp a) = s_0^{\mp} \tag{9.15}$$

$$s_{2k}(\mp a) = s_{2k}^{\mp}, \quad hs'_{2k}(\mp a) = d_{2k}^{\mp}, \quad k \geq 1 \tag{9.16}$$

where we have separated conditions (9.15) on the principal stress from conditions

(9.16) on the residual stresses.

The integration of equation (9.6) subject to condition (9.15) yields the compatibility conditions

$$2h(\overset{+}{s}_0 - \overset{-}{s}_0) = -\int_{-a}^{a} p_x dx \tag{9.17}$$

necessary for overall static equilibrium, together with the solution

$$2hs_0 = h(\overset{+}{s}_0 + \overset{-}{s}_0) + \frac{1}{2}\int_{-a}^{a} p_x dx - \int_{-a}^{x} p_x dx \tag{9.18}$$

for the principal stress. Relations (9.17) and (9.18) correspond respectively to relations (5.11) and (5.12*) in the general case.

With the determination of the principal stress thus completed, the next step in the general case would be the introduction of this explicit form for s_0 into the formulae for the stress coefficients. However, in the restricted case, we can proceed more directly. We use the differential equation (9.6) to eliminate s_0'' from relations (9.11) and thereby obtain for the transverse normal stress coefficients $\{\sigma_{2k},\ k \geq 1\}$, expressions that involve only the residual axial stress coefficients $\{s_{2k},\ k \geq 1\}$ and the applied forces: separating the formula for σ_0 from the expressions for the higher coefficients, we find

$$\sigma_0 = \frac{1}{15} h^2 s_2'' + \frac{1}{2} p^* + \frac{1}{6} hp_x' \tag{9.19}$$

and

$$\sigma_2 = h^2 \left[-\frac{2}{21} s_2'' + \frac{1}{63} s_4'' \right] - \frac{1}{6} hp_x' \tag{9.20a}$$

$$\sigma_{2k} = h^2 \left[\frac{s_{2k-2}''}{(4k-3)(4k-1)} - \frac{2s_{2k}''}{(4k-1)(4k+3)} + \frac{s_{2k+2}''}{(4k+3)(4k+5)} \right], \quad k \geq 2 \tag{9.20b}$$

In passing, we note that, in the sequence of relations (9.20), inhomogeneous terms appear only in the first relation, namely in the expression for σ_2: their absence in the expressions for the coefficients of higher index is due to the fact that we have neglected all body force effects associated with the problem of stretching.

An inspection of relations (9.19) and (9.20a) immediately gives the resolution of σ_0 and σ_2 into their principal and residual components: in accordance with (8.8) and (8.10), we have

$$\sigma_0 = \overset{P}{\sigma}_0 + \overset{R}{\sigma}_0, \qquad \sigma_2 = \overset{P}{\sigma}_2 + \overset{R}{\sigma}_2 \tag{9.21a,b}$$

where

$$\overset{P}{\sigma_0} = \frac{1}{2}p^* + \frac{1}{6}hp'_x, \qquad \overset{R}{\sigma_0} = \frac{1}{15}h^2 s_2'' \tag{9.22a,b}$$

and

$$\overset{P}{\sigma_2} = -\frac{1}{6}hp'_x, \qquad \overset{R}{\sigma_2} = -\frac{2}{21}h^2 s_2'' + \frac{1}{63}h^2 s_4'' \tag{9.23a,b}$$

In the notation of the infinite dimensional linear vector space, induced by the expansion (9.5c) in terms of Legendre polynomials of even index, we write

$$\sigma = \overset{P}{\sigma} + \overset{R}{\sigma} \tag{9.24}$$

where

$$\overset{P}{\sigma} = \{\overset{P}{\sigma_0}, \overset{P}{\sigma_2}, 0, 0, \ldots, 0, \ldots \quad\} \tag{9.25a}$$

$$\overset{R}{\sigma} = \{\overset{R}{\sigma_0}, \overset{R}{\sigma_2}, \sigma_4, \sigma_6, \ldots, \sigma_{2k}, \ldots\} \tag{9.25b}$$

This decomposition corresponds to that made in (8.12) and (8.13).

We next consider the representations for the displacement quantities. For the transverse displacement and the rotation, we have the respective expansions

$$w = \sum_{k=0}^{\infty} w_{2k+1}(x)P_{2k+1}(t) \tag{9.26a}$$

$$\omega = \sum_{k=0}^{\infty} \omega_{2k+1}(x)P_{2k+1}(t) \tag{9.26b}$$

whose coefficients are determined in terms of the stress coefficients. The transverse displacement coefficients in the expansion (9.26a) are given by

$$w_{2k+1} = \frac{\mathrm{h}}{E}\left[\frac{\nu s_{2k+2} - \sigma_{2k+2}}{4k+5} - \frac{\nu s_{2k} - \sigma_{2k}}{4k+1}\right], \quad k \geq 0 \tag{9.27}$$

which follows from (6.3) by setting $n = 2k+1$: the corresponding expressions for the rotation coefficients in the expansion (9.26b) are contained in the restricted form of (6.14b*) with $n = 2k+1$, namely

$$\omega_{2k+1} = \frac{\mathrm{h}}{E}\left[\frac{s'_{2k+2} + \sigma'_{2k+2}}{4k+5} - \frac{s'_{2k} + \sigma'_{2k}}{4k+1}\right], \quad k \geq 1 \tag{9.28}$$

Similarly the higher coefficients in the expansion for the axial displacement

$$u = \sum_{k=0}^{\infty} u_{2k}(x) P_{2k}(t) \tag{9.29}$$

are given by the restricted form of relations (6.17b*) with even index, namely

$$u_{2k} = -\frac{h^2}{E}\left[\frac{(2+\nu)s'_{2k-2}+\sigma'_{2k-2}}{(4k-3)(4k-1)} - 2\frac{(2+\nu)s'_{2k}+\sigma'_{2k}}{(4k-1)(4k+3)} + \frac{(2+\nu)s'_{2k+2}+\sigma'_{2k+2}}{(4k+3)(4k+5)}\right],$$

$$k \geq 1 \tag{9.30}$$

while the zero-th coefficient, representing the mean axial displacement, satisfies the restricted form of equation (7.10a). If we note the restriction (9.8) in connection with formula (7.7a), we see that

$$\phi_0 = 0 \tag{9.31}$$

The latter, together with formula (9.19) for σ_0, when introduced into (7.10a) gives the equation satisfied by u_0 in the restricted case, namely

$$Eu'_0 = s_0 - \nu[\frac{1}{2}p^* + \frac{1}{6}hp'_x + \frac{1}{15}h^2 s''_2] \tag{9.32a}$$

which is to be solved subject to the condition (7.14), namely

$$u_0(0) = -\sum_{k=1}^{\infty}(-1)^k \frac{(2k-1)!!}{(2k)!!} u_{2k}(0) \tag{9.32b}$$

Following the resolution of σ_0 made in (9.22), we now make the corresponding decomposition of u_0 in accordance with the procedure prescribed in (8.20) to (8.22). Setting

$$u_0 = U + U_R \tag{9.33}$$

we have for the determination of the residual component

$$EU'_R = -\frac{1}{15}\nu h^2 s''_2, \quad U_R(0) = U_{R0} \tag{9.34a,b}$$

where*

$$U_{R0} = -\sum_{k=1}^{\infty}(-1)^k \frac{(2k-1)!!}{(2k)!!} u^R_{2k}(0) \tag{9.34c}$$

* The decomposition of the higher coefficients u_{2k} in the form $u_{2k} = u^P_{2k} + u^R_{2k}$ has been outlined in Section Eight for the general case: for the restricted case, the resolution follows the simpler pattern given above for the coefficients σ_{2k}.

while the principal component satisfies

$$EU' = s_0 - \nu(\frac{1}{2}p^* + \frac{1}{6}hp_x') \tag{9.35a}$$

subject to

$$U(0) = U_0 \tag{9.35b}$$

in which

$$U_0 = -\sum_{k=1}^{6}(-1)^k\frac{(2k-1)!!}{(2k)!!}u_{2k}^P(0) \tag{9.36}$$

where we recall relations (8.19a). Substituting for s_0 from (9.18) into (9.35a), we integrate, and noting (9.35b), we obtain for the principal component of the mean axial displacement

$$U = U_0 + \frac{1}{2E}\left[[s_0^+ + s_0^- + \frac{1}{2h}\int_{-a}^{a}p_x dx]x - \frac{1}{h}\int_0^x[\int_{-a}^{x}p_x dx]dx \right.$$
$$\left. - \nu\int_0^x p^* dx - \frac{1}{3}\nu h[p_x(x) - p_x(0)]\right] \tag{9.37}$$

From (9.34), we have for the residual component of the mean axial displacement

$$U_R = U_{R0} - \frac{1}{15}\frac{\nu}{E}h^2[s_2'(x) - s_2'(0)] \tag{9.38}$$

The expression for the mean displacement u_0 then follows from the addition of formulae (9.37) and (9.38).

It remains to investigate the residual axial stresses.

The non-vanishing elements in the sequence $\{\phi_n\}$ are given, in terms of the transverse normal stress coefficients, by the restriced form of (7.7c) with even index, namely

$$\phi_{2k} = h^2\left[\frac{\sigma_{2k-2}''}{(4k-3)(4k-1)} - \frac{2\sigma_{2n}''}{(4k-1)(4k+3)} + \frac{\sigma_{2k+2}''}{(4k+3)(4k+5)}\right], \quad k \geq 1 \tag{9.39}$$

In this system of relations, the quantity σ_0 occurs only in the case $k = 1$, namely in the expression for ϕ_2: writing this case separately and substituting for σ_0 from (9.19), we have the alternative form

$$\phi_2 = h^2[-\frac{2}{21}\sigma_2'' + \frac{1}{63}\sigma_4''] + \frac{1}{45}h^4 s_2'''' + \frac{1}{6}h^2 p^{*\,\prime\prime} + \frac{1}{18}h^3 p_x''' \tag{9.40a}$$

$$\phi_{2k} = h^2\left[\frac{\sigma_{2k-2}''}{(4k-3)(4k-1)} - \frac{2\sigma_{2k}''}{(4k-1)(4k+3)} + \frac{\sigma_{2k+2}''}{(4k+3)(4k+5)}\right],\ k\geq 2 \tag{9.40b}$$

With the quantities $\{\phi_{2k},\ k\geq 1\}$ given by (9.40) and the quantities $\{\sigma_{2k},\ k\geq 1\}$ given by (9.20), it is now evident that relations (7.10c) with even index, namely,

$$\phi_{2k} + 2\sigma_{2k} + s_{2k} = 0,\quad k\geq 1 \tag{9.41}$$

constitute a self-contained system of differential equations for the residual axial stresses, to be solved subject to the boundary conditions (9.16).

The next step is the introduction of notation that allows relations (9.20), (9.40) and (9.41), together with the boundary conditions (9.16), to be written in compact vector form.

The infinite dimensional vector, whose components are the residual axial stress coefficients with even index is denoted by S_e: the k-th element of S_e being s_{2k}, we have

$$S_e = \{s_{2k}, 1\leq k\leq\infty\} = \{s_2, s_4, s_6, \ldots, s_{2k}, \ldots\} \tag{9.42}$$

Similarly Σ_e represents the vector whose components are the corresponding normal stress coefficients, namely

$$\Sigma_e = \{\sigma_{2k}, 1\leq k\leq\infty\} = \{\sigma_2, \sigma_4, \sigma_6, \ldots, \sigma_{2k}, \ldots\} \tag{9.43}$$

If we let Φ_e designate the vector consisting of the corresponding elements in the sequence $\{\phi_{2k}\}$ so that

$$\Phi_e = \{\phi_{2k}, 1\leq k\leq\infty\} = \{\phi_2, \phi_4, \phi_6, \ldots, \phi_{2k}, \ldots\} \tag{9.44}$$

then the system of equations (9.41) takes the vector form

$$\Phi_e + 2\Sigma_e + S_e = 0 \tag{9.45}$$

If, in addition, we introduce the vectors $S_e^{\mp}$ and $D_e^{\mp}$ by setting

$$S_e^{\mp} = \{s_{2k}^{\mp},\ 1\leq k\leq\infty\} = \{s_2^{\mp}, s_4^{\mp}, s_6^{\mp}, \ldots, s_{2k}^{\mp}, \ldots\} \tag{9.46a}$$

$$D_e^{\mp} = \{d_{2k}^{\mp},\ 1 \le k \le \infty\} = \{d_2^{\mp},\ d_4^{\mp},\ d_6^{\mp}, \ldots, d_{2k}^{\mp}, \ldots\} \tag{9.46b}$$

then the boundary conditions (9.16) may be written

$$x = \mp a: \quad S_e = S_e^{\mp}, \quad hS_e' = D_e^{\mp} \tag{9.47}$$

complementing the vector differential equation (9.45).

We may view relations (9.20) and (9.40) as linear transformations of the sequence space into itself. In the vector formulation of these transformations the inhomogeneous terms, due to the applied axial and transverse forces, will be furnished respectively by the vectors

$$F_e^{(A)} = \{hp_x',\ 0,\ 0, \ldots.\}, \quad F_e^{(T)} = \{p^*,\ 0,\ 0, \ldots.\} \tag{9.48}$$

in each of which only the first element is nonzero. The linear operator of the transformation (9.20) relating S_e to Σ_e, which also occurs in the system (9.40) relating Σ_e to Φ_e may be defined in terms of the one-sided infinite tridiagonal matrix M_e, where*

$$M_e = [m_{k,j}^{(e)}]_1^{\infty} \tag{9.49}$$

in which both suffices k and j range from one to infinity. The non-zero elements are given by

$$m_{k,k}^{(e)} = \frac{2}{(4k-1)(4k+3)}, \quad m_{k,k+1}^{(e)} = -\frac{1}{(4k+3)(4k+5)}, \quad 1 \le k \le \infty \tag{9.50a,b}$$

$$m_{k,k-1}^{(e)} = -\frac{1}{(4k-3)(4k-1)}, \quad 2 \le k \le \infty \tag{9.50c}$$

while all other elements vanish, namely

$$m_{k,j}^{(e)} = 0 \qquad \begin{matrix} 1 \le j \le k-2, & k \ge 3 \\ \\ k+2 \le j \le \infty, & k \ge 1 \end{matrix} \tag{9.50d}$$

From the matrix M_e, we form the matrix function $L_e(\lambda)$ by scalar multiplication

* In a more general context M_e is a one-sided infinite band matrix of order one.

by the parameter λ,

$$L_e(\lambda) = \lambda M_e = [m^{(e)}_{k,j}\lambda]_1^\infty \tag{9.51}$$

from which we shall form the differential operator by direct substitution. In the system (9.40) there occurs a second factor which is most conveniently given in terms of a second matrix function $J(\lambda)$ formed from the basis element matrix* E_{11} by scalar multiplication by λ. Thus with

$$E_{11} = [\delta_{k1}\delta_{j1}]_1^\infty \tag{9.52}$$

the matrix function $J(\lambda)$ is given by

$$J(\lambda) = \lambda E_{11} = [\delta_{k1}\delta_{j1}\lambda]_1^\infty \tag{9.53}$$

where δ_{k1} and δ_{j1} are the Kronecker deltas.

We now introduce the thickness scaled variable ξ and the associated parameter α measuring the length-thickness ratio by setting

$$\xi = \frac{x}{h}, \quad \alpha = \frac{a}{h} \tag{9.54}$$

so that in terms of these dimensionless quantities the edges are specified by

$$\xi = \mp\alpha \tag{9.55}$$

and the vector S_e representing the residual axial stresses is to be considered an unknown vector function of ξ to be determined over the interval

$$-\alpha \leq \xi \leq \alpha \tag{9.56}$$

If, in terms of the derivative element for the independent variable ξ, namely

$$D = \frac{d}{d\xi} = h\frac{d}{dx} \tag{9.57}$$

we construct the matrix differential operators $L_e(D^2)$ and $J(D^2)$ by appropriate substitution in the matrix functions of (9.51) and (9.53), namely

$$L_e(D^2) = [m^{(e)}_{k,j}D^2]_1^\infty, \quad J(D^2) = [\delta_{k1}\delta_{j1}D^2]_1^\infty \tag{9.58}$$

then, in vector form, the transformations (9.20) reads

$$\Sigma_e = -L_e(D^2)S_e - \frac{1}{6}F_e^{(A)} \tag{9.59}$$

* The basis element E_{kj} has entry 1 in the $k-j$ position and zeros elsewhere.

while the system (9.40) takes the form

$$\Phi_e = -L_e(D^2)\Sigma_e + \frac{1}{45}[J(D^2)]^2 S_e + \frac{1}{6}D^2[F_e^{(T)} + \frac{1}{3}F_e^{(A)}] \tag{9.60}$$

To obtain the expression for Φ_e directly in terms of S_e, we introduce Σ_e from (9.59) into (9.60): noting that

$$L_e(D^2)F_e^{(A)} = \frac{2}{21}D^2F_e^{(A)} \tag{9.61}$$

we find

$$\Phi_e = \{[L_e(D^2)]^2 + \frac{1}{45}[J(D^2)]^2\} S_e + D^2[\frac{1}{6}F_e^{(T)} + \frac{1}{14}F_e^{(A)}] \tag{9.62}$$

The explicit form for the vector differential equation satisfied by S_e now follows from the introduction of Σ_e from (9.59) and Φ_e from (9.62) into (9.45): if we also rewrite the boundary conditions (9.47) in the normalized notation of (9.54) to (9.57), then the boundary value problem for the residual stresses may be stated as

$$\{[L_e(D^2)]^2 + \frac{1}{45}[J(D^2)]^2 - 2L_e(D^2) + I\}S_e = \frac{1}{3}F_e^{(A)} - D^2[\frac{1}{6}F_e^{(T)} + \frac{1}{14}F_e^{(A)}] \tag{9.63a}$$

$$S_e(\mp\alpha) = S_e, \quad DS_e(\mp\alpha) = D_e^{\mp} \tag{9.63b}$$

in which we have used I to denote the one-sided infinite unit matrix.

The solution of the inhomogeneous system (9.63a) can, by the standard variation-of-constants procedure, be generated from the solution of the associated homogeneous system: it suffices therefore to make a detailed analysis of the latter. Moreover, the vanishing of the forcing terms is associated with a constant axial traction p_x and a "pinching" force p^* that has a linear axial variation. Accordingly, we take

$$p_x' = 0, \quad p^{*\prime\prime} = 0 \tag{9.64}$$

so that

$$F_e^{(A)} = 0, \quad D^2F_e^{(T)} = 0 \tag{9.65}$$

and equation (9.63a) assumes the homogeneous form

$$\boxed{\{[L_e(D^2) - I]^2 + \frac{1}{45}[J(D^2)]^2\} S_e = 0} \tag{9.66}$$

where we have rearranged the terms in the differential operator.

Although we have not succeeded in determining the factors explicitly, it should be possible to effect a decomposition of the above operator in the form of a product of a pair of mutually conjugate operators. The system of equations (9.66) could then be formally written as

$$[\mathcal{L}_e(D^2) - I][\mathcal{L}_e^*(D^2) - I]\, S_e = 0 \tag{9.67}$$

so that the problem would reduce to the determination of the general solution for either of the systems

$$[\mathcal{L}_e(D^2) - I]X = 0, \quad [\mathcal{L}_e^*(D^2) - I]\, X = 0 \tag{9.68a,b}$$

An inversion would transform these to the equivalent form

$$X'' = AX, \quad X'' = A^*X \tag{9.69a,b}$$

for which the formal construction of the four independent fundamental matrices would be straightforward. If we were to postmultiply each of the fundamental matrices associated with (9.69a) by an arbitrary vector and form the sum, we could then complete the general solution by adding these to the pair of conjugate vectors. The real and imaginary parts of the arbitrary vectors would then constitute the four unknown vectors to be determined from the four boundary conditions (9.63b).

The validity of such a formal solution is made evident if we observe that in the matrix M_e, the nonvanishing elements in the kth, row or column are of order $1/k^2$. This guarantees complete continuity for the operator in (9.66) from which it follows that the associated spectrum is discrete. In fact, the matrices constructed from the associated eigenvectors give an alternative construction for the general solution. Moreover, from the spectral equation

$$\det\{[L_e(\lambda^2) - I]^2 + \tfrac{1}{45}[J(\lambda^2)]^2\} = 0 \tag{9.70}$$

it is evident that the characteristic values occur in groups of four in the form

$$\mp\lambda_i, \quad \mp\bar{\lambda}_i \quad 1 \le i \le \infty \tag{9.71}$$

In the case of the semi-infinite strip, the regularity conditions immediately exclude those roots with positive real part and we find that the roots with negative real part then directly reflect the exponents of decay for the residual stresses.

The form (9.71) for the spectrum would tend to confirm the conjectured decomposition (9.67) for the operator.

10. The Problem of Bending: Restricted Case

In the problem of bending, the surface forces are distributed so that

$$p^* = 0, \quad p_x = 0 \qquad \textbf{(10.1a,b)}$$

while the specification of the edge stress vectors is such that, in their dependence on the thickness coordinate, the normal components are odd while the shear components are even: hence,

$$s_{2k}^{\mp} = 0, \quad \tau_{2k+1}^{\mp} = 0, \quad k \geq 0 \qquad \textbf{(10.2)}$$

which, when combined with (10.1b) and (5.8a), is equivalent to

$$s_{2k}^{\mp} = 0, \quad k \geq 0: \quad d_{2k}^{\mp} = 0, \quad k \geq 1 \qquad \textbf{(10.3)}$$

Thus, the coefficients associated with the stretching effects satisfy homogeneous differential equations with homogeneous boundary conditions and therefore, vanish identically: explicitly, we have

$$s_{2k}(x) \equiv 0, \quad \sigma_{2k}(x) \equiv 0, \quad u_{2k}(x) \equiv 0, \; k \geq 0 \,. \qquad \textbf{(10.4a)}$$

$$\tau_{2k+1}(x) \equiv 0, \; w_{2k+1}(x) \equiv 0, \; \omega_{2k+1}(x) \equiv 0, \; k \geq 0 \,. \qquad \textbf{(10.4b)}$$

$$\phi_{2k}(x) \equiv 0, \quad k \geq 0 \,. \qquad \textbf{(10.4c)}$$

Accordingly, the expansions for the stress components take the form

$$\sigma_x = \sum_{k=0}^{\infty} s_{2k+1}(x) P_{2k+1}(t) \qquad \textbf{(10.5a)}$$

$$\tau = \sum_{k=0}^{\infty} \tau_{2k}(x) P_{2k}(t) \qquad \textbf{(10.5b)}$$

$$\sigma = \sum_{k=0}^{\infty} \sigma_{2k+1}(x) P_{2k+1}(t) \qquad \textbf{(10.5c)}$$

The zero-th coefficients in the first two expansions, namely the principal stresses s_1 and τ_0, satisfy the pair of equations

$$\frac{2}{3} h^2 s_1' - 2h\tau_0 + hp_x^* = 0 \qquad \textbf{(10.6)}$$

$$2h\tau_0' + p + 2hf_0 = 0 \qquad \textbf{(10.7)}$$

which follows from the introduction of the restriction

$$h' = 0 \tag{10.8}$$

into equations (4.13) and (4.14), respectively. The factor τ_0 may be eliminated by the introduction of τ_0' from (10.7) into the differentiated form of (10.6) yielding the equation

$$\frac{2}{3} h^2 s_1'' + p + 2hf_0 + hp_x^{*\prime} = 0 \tag{10.6$'$}$$

for the determination of the principal axial stress coefficients s_1: the equations satisfied by the residual axial stress coefficients $\{s_{2k+1}, k \geq 1\}$ in the expansion (10.5a) will be stated later.

The shear stress coefficients in the expansions (10.5b) are given in terms of the axial stress coefficients by the relations

$$\tau_0 = \frac{1}{3} hs_1' + \frac{1}{2} p_x^* \tag{10.9a}$$

$$\tau_{2k} = h \left[\frac{s_{2k+1}'}{4k+3} - \frac{s_{2k-1}'}{4k-1} \right], \quad k \geq 1 \tag{10.9b}$$

The latter follows from the introduction of (10.8) into relations (3.10) with $n = 2k$, while the former is a rearrangement of (10.6).

We may write the transverse normal stress coefficients in the expression (10.5c) both in terms of the axial stress coefficients and in terms of the shear stress coefficients. For the latter, we have the restricted form ($h' = 0$) of relation (3.13) with $n = 2k+1$, namely

$$\sigma_{2k+1} = h \left[\frac{\tau_{2k+2}'}{4k+5} - \frac{\tau_{2k}'}{4k+1} \right], \quad k \geq 0 \tag{10.10}$$

The corresponding expressions in terms of the axial coefficients follow from the introduction of relations (10.9) into relations (10.10): thus for $k = 0$, the latter yields the restricted form of (4.20b), while for $k \geq 1$, it becomes the restricted form of relations (3.14) with odd index: we obtain

$$\sigma_1 = h^2 [-\frac{2}{5} s_1'' + \frac{1}{35} s_3''] - \frac{1}{2} hp_x^{*\prime} \tag{10.11a}$$

$$\sigma_{2k+1} = h^2 \left[\frac{s_{2k-1}''}{(4k-1)(4k+1)} - \frac{2s_{2k+1}''}{(4k+1)(4k+5)} + \frac{s_{2k+3}''}{(4k+5)(4k+7)} \right], \quad k \geq 1 \tag{10.11b}$$

Writing relation (10.9) for successive values of k, we sum the series, and, noting the cancellations, we obtain the restricted form of (5.5), namely

$$hs'_{2k+1} = (4k+3)[\sum_{j=0}^{k} \tau_{2j} - \frac{1}{2}p_x^*], \quad k \geq 0 \tag{10.12}$$

from which, if we set $k = 0$, we retrieve relation (10.6). Thereby, the quantities $d_{2k+1}^{\mp}$ defined by

$$d_{2k+1}^{\mp} = (4k+3)[\sum_{j=0}^{k} \tau_{2j}^{\mp} - \frac{1}{2}p_x^*(\mp a)], \quad k \geq 0 \tag{10.13}$$

enable us to transform the edge conditions

$$s_{2k+1}(\mp a) = s_{2k+1}^{\mp}, \quad \tau_{2k}(\mp a) = \tau_{2k}^{\mp}, \quad k \geq 0 \tag{10.14}$$

to the equivalent system

$$s_1(\mp a) = s_1^{\mp}, \quad \tau_0(\mp a) = \tau_0^{\mp} \tag{10.15a,b}$$

$$s_{2k+1}(\mp a) = s_{2k+1}^{\mp}, \quad hs'_{2k+1}(\mp a) = d_{2k+1}^{\mp}, \quad k \geq 1 \tag{10.16}$$

where we have separated conditions (10.15) on the principal stresses from conditions (10.16) on the residual stresses.

Since both h and f_0 are constant, the quantity $f_0^*(x)$, corresponding to that defined in (5.13), takes the simpler form

$$f_0^*(x) = 2h(x+a)f_0 + \int_{-a}^{x} p\,dx \tag{10.17a}$$

so that

$$f_0^*(a) = 4ahf_0 + \int_{-a}^{a} p\,dx \tag{10.17b}$$

Then the integration of equation (10.7), subject to conditions (10.15b), yields the first compatibility condition

$$2h(\overset{+}{\tau}_0 - \overset{-}{\tau}_0) = -f_0^*(a) = -4ahf_0 - \int_{-a}^{a} p\,dx \tag{10.18}$$

necessary for overall static equilibrium, together with the solution

$$2h\tau_0 = h(\overset{+}{\tau_0} + \overset{-}{\tau_0}) + \frac{1}{2} f_0^*(a) - f_0^*(x)$$

$$= h(\overset{+}{\tau_0} + \overset{-}{\tau_0}) - 2xhf_0 + \frac{1}{2}\int_{-a}^{a} p\,dx - \int_{-a}^{x} p\,dx \tag{10.19}$$

for the principal shear stress. With τ_0 thus determined, the integration of equation (10.6) subject to conditions (10.15a) gives the second static equilibrium condition

$$\frac{2}{3}h^2(\overset{+}{s_1} - \overset{-}{s_1}) = 2h\int_{-a}^{a} \tau_0 dx - h\int_{-a}^{a} p_x^* dx$$

$$= 2a[h(\overset{+}{\tau_0} + \overset{-}{\tau_0}) + \frac{1}{2} f_0^*(a)] - \int_{-a}^{a} [f_0^*(x) + hp_x^*(x)]dx \tag{10.20}$$

together with the solution

$$\frac{2}{3}h^2 s_1 = \frac{2}{3}h^2\overset{-}{s_1} + 2h\int_{-a}^{x} \tau_0 dx - h\int_{-a}^{x} p_x^* dx$$

$$= \frac{2}{3}h^2\overset{-}{s_1} + (x+a)[h(\overset{+}{\tau_0} + \overset{-}{\tau_0}) + \frac{1}{2} f_0^*(a)] - \int_{-a}^{x} [f_0^*(x) + hp_x^*(x)]dx \tag{10.21}$$

for the principal axial stress. Relations (10.18) to (10.21) correspond to relations (5.14) to (5.17) in the general case.

We now use equation (10.6′) to substitute for s_1'' in relations (10.11) and thereby, for the transverse normal stress coefficients $\{\sigma_{2k+1},\ k \geq 0\}$, we obtain expressions that involve only the residual axial stress coefficients $\{s_{2k+1},\ k \geq 0\}$ and the applied forces: separating the formula for σ_1 from the expressions for the higher coefficients, we find

$$\sigma_1 = \frac{1}{35}h^2 s_3'' + \frac{3}{5}(p + 2hf_0) + \frac{1}{10}hp_x^{*\prime} \tag{10.22}$$

and

$$\sigma_3 = h^2[-\frac{1}{45}s_3'' + \frac{1}{99}s_5''] - \frac{1}{10}(p + 2hf_0) - \frac{1}{10}hp_x^{*\prime} \tag{10.23a}$$

$$\sigma_{2k+1} = h^2\left[\frac{s''_{2k-1}}{(4k-1)(4k+1)} - \frac{2s''_{2k+1}}{(4k+1)(4k+5)} + \frac{s''_{2k+3}}{(4k+5)(4k+7)}\right], \quad k \geq 2 \tag{10.23b}$$

An inspection of relations (10.22) and (10.23a) immediately gives the resolution of σ_1 and σ_3 into their principal and residual components: in accordance with (8.9) and (8.11), we have

$$\sigma_1 = \overset{P}{\sigma_1} + \overset{R}{\sigma_1}, \quad \sigma_3 = \overset{P}{\sigma_3} + \overset{R}{\sigma_3} \tag{10.24}$$

where

$$\overset{P}{\sigma_1} = \frac{3}{5}(p + 2hf_0) + \frac{1}{10}hp_x^{*\prime}, \quad \overset{R}{\sigma_1} = \frac{1}{35}h^2 s_3'' \tag{10.25a,b}$$

and

$$\overset{P}{\sigma_3} = -\frac{1}{10}(p + 2hf_0 + p_x^{*\prime}), \quad \overset{R}{\sigma_3} = h^2[-\frac{2}{45}s_3'' + \frac{1}{99}s_5''] \tag{10.26a,b}$$

In the notation of the infinite dimensional linear vector space induced by the expansion (10.5c) in terms of Legendre polynomials of odd index, we write

$$\sigma = \overset{P}{\sigma} + \overset{R}{\sigma} \tag{10.27}$$

where

$$\overset{P}{\sigma} = \{\overset{P}{\sigma_1}, \overset{P}{\sigma_3}, 0, 0, \ldots, 0, \ldots\} \tag{10.28a}$$

$$\overset{R}{\sigma} = \{\overset{R}{\sigma_1}, \overset{R}{\sigma_3}, \sigma_5, \sigma_7, \ldots, \sigma_{2k+1}, \ldots\} \tag{10.28b}$$

This composition corresponds to that made in (8.9) and (8.11).

In noting the form taken by the representations for the displacement quantities, we start with the transverse displacement: the zero-th coefficient in the expansion

$$w = \sum_{k=0}^{\infty} w_{2k}(x)P_{2k}(t) \tag{10.29}$$

will be determined presently while the higher coefficients are given by relations (6.3) with $n = 2k$, namely

$$\begin{aligned} w_2 &= \frac{h}{E}\left[\frac{\nu s_3 - \sigma_3}{7} - \frac{\nu s_1 - \sigma_1}{3}\right] - \frac{1}{3}\frac{h^2}{E}f_0 \\ w_{2k} &= \frac{h}{E}\left[\frac{\nu s_{2k+1} - \sigma_{2k+1}}{4k+3} - \frac{\nu s_{2k-1} - \sigma_{2k-1}}{4k-1}\right], \quad k \geq 2 \end{aligned} \tag{10.30}$$

The coefficients in the corresponding representation for the rotation

$$\omega = \sum_{k=0}^{\infty} \omega_{2k}(x)P_{2k}(t) \tag{10.31}$$

are given by the restricted form of either (6.14) or (6.14*) with $n = 2k$, namely

$$\omega_0 = \frac{(1+\nu)}{E}\tau_0 - w_0' = \frac{(1+\nu)}{E}[\frac{1}{3}hs_1' + \frac{1}{2}p_x^*] - w_0' \tag{10.32a}$$

$$\omega_{2k} = \frac{h}{E}\left[\frac{s_{2k+1}' + \sigma_{2k+1}'}{4k+3} - \frac{s_{2k-1}' + \sigma_{2k-1}'}{4k-1}\right], \quad k \geq 1 \tag{10.32b}$$

For the axial displacement, we have the expansion

$$u = \sum_{k=0}^{\infty} u_{2k+1}(x)P_{2k+1}(t) \tag{10.33}$$

where the coefficients are given by the formulae

$$u_1 = \frac{h^2}{E}\left[\frac{(12+11\nu)s_1' + \sigma_1'}{15} - \frac{(2+\nu)s_3' + \sigma_3'}{35}\right] + \frac{1}{2}\frac{h}{E}p_x^* - hw_0' \tag{10.34a}$$

$$u_{2k+1} = \frac{h^2}{E}\left[\frac{(2+\nu)s_{2k-1}' + \sigma_{2k-1}'}{(4k-1)(4k+1)} - 2\frac{(2+\nu)s_{2k+1}' + \sigma_{2k+1}'}{(4k+1)(4k+5)} + \frac{(2+\nu)s_{2k+3}' + \sigma_{2k+3}'}{(4k+5)(4k+7)}\right], \quad k \geq 1 \tag{10.34b}$$

obtained from taking relations (6.17*) with even index and introducing the restriction (10.8).

The zero-th coefficient in (10.29) representing the mean transverse displacement satisfies the restricted form of (7.10b). If we note the restriction (10.8) in connection with (7.7b), we see that

$$\phi_1 = \nu[-\frac{1}{3}h^2 s_1'' - \frac{1}{2}hp_x^{*\prime}] - \frac{h^2}{5}[\frac{\sigma_1}{3} - \frac{\sigma_3}{7}]'' \tag{10.35}$$

In relation (10.35), we now substitute for s_1'' from (10.6′) and also introduce σ_1 and σ_3 from relations (10.22) and (10.23a) respectively: if we rearrange and recall that we have already assumed that f_0 is constant, we find

$$\phi_1 = \frac{1}{2}\nu(p+2hf_0) - \frac{3}{70}h^2(p + \frac{2}{9}hp_x^{*\prime})'' - \frac{1}{315}h^4(s_3'' - \frac{1}{11}s_5'')'' \tag{10.36}$$

so that in accordance with (8.16) and (8.17), we may set

$$\phi_1 = \phi_1^P + \phi_1^R \tag{10.37}$$

where

$$\phi_1^P = \frac{1}{2}\nu(p+2hf_0) + \frac{3}{70}h^2(p + \frac{2}{9}hp_x^{*\prime})'' \tag{10.38a}$$

$$\phi_1^R = -\frac{1}{315}h^4(s_3'' - \frac{1}{11}s_5'')'' \tag{10.38b}$$

The equation for the determination of w_0 now follows from the introduction of σ_1 from (10.22) and ϕ_1 from (10.36) into (7.10b), yielding the restricted form of the latter equation, which, after some rearrangement, reads

$$-Ehw_0'' = s_1 + (\frac{6}{5} + \frac{1}{2}\nu)p + 2(\frac{6}{5}+\nu)hf_0 + \frac{1}{5}hp_x^{*\prime} - \frac{3}{70}h^2(p+\frac{2}{9}hp_x^{*\prime})''$$

$$+\frac{2}{35}h^2[s_3 - \frac{1}{18}h^2(s_3''-\frac{1}{11}s_5'')]'' \tag{10.39}$$

This equation is to be solved subject to the conditions (7.15), namely

$$w_0(0) = -\sum_{k=1}^{\infty}(-1)^k\frac{(2k-1)!!}{(2k)!!}w_{2k}(0), \tag{10.40}$$

$$\omega_0(0) = -\sum_{k=1}^{\infty}(-1)^k\frac{(2k-1)!!}{(2k)!!}\omega_{2k}(0) \tag{10.41}$$

whence, if we note relation (10.32a) in connection with the latter condition (10.41), we have the alternate form

$$w_0'(0) = \bar{\omega}_0 + \sum_{k=1}^{\infty}(-1)^k\frac{(2k-1)!!}{(2k)!!}\omega_{2k}(0) \tag{10.41*}$$

where, noting (10.19)

$$\bar{\omega}_0 = \frac{(1+\nu)}{E}\tau(0) = \frac{(1+\nu)}{2E}[\overset{+}{\tau}_0 + \overset{-}{\tau}_0 + \frac{1}{2h}(\int_0^a pdx - \int_{-a}^0 pdx)] \tag{10.42}$$

Following the resolution of σ_1 in (10.22) and of ϕ_1 in (10.38), we now make a corresponding decomposition of w_0 in accordance with the procedure prescribed in (8.24) and (8.25). Setting*

$$w_0 = W + W_R \tag{10.43}$$

we have for the determination of the residual component

$$-EhW_R'' = \frac{2}{35}h^2[s_3 - \frac{1}{18}h^2(s_3'' - \frac{1}{11}s_5'')]'' \tag{10.44a}$$

with

$$W_R(0) = W_{R0}, \quad W_R'(0) = W_{R0}' \tag{10.44b}$$

where

$$W_{R0}(0) = -\sum_{k=1}^{\infty}(-1)^k\frac{(2k-1)!!}{(2k)!!}\overset{R}{w}_{2k}(0), \tag{10.44c}$$

$$W_{R0}'(0) = +\sum_{k=1}^{\infty}(-1)^k\frac{(2k-1)!!}{(2k)!!}\overset{R}{\omega}_{2k}(0) \tag{10.44c*}$$

Hence, the principal component satisfies

$$-EhW'' = s_1 + (\frac{6}{5}+\frac{1}{2}\nu)p + 2(\frac{6}{5}+\nu)hf_0 + \frac{1}{5}hp_x^{*\prime} - \frac{3}{70}h^2(p + \frac{2}{9}hp_x^{*\prime})'' \tag{10.45a}$$

subject to the conditions

$$W(0) = W_0, \quad W'(0) = W_0' \tag{10.45b}$$

in which

$$W_0 = -\sum_{k=1}^{5}(-1)^k\frac{(2k-1)!!}{(2k)!!}\overset{P}{w}_{2k}(0), \tag{10.45c}$$

$$W_0' = \overline{\omega}_0 + \sum_{k=1}^{5}(-1)^k\frac{(2k-1)!!}{(2k)!!}\overset{P}{\omega}_{2k}(0) \tag{10.45d}$$

* The decomposition of the higher coefficients, namely

$$w_{2k} = \overset{P}{w}_{2k} + \overset{R}{w}_{2k}, \quad \omega_{2k} = \overset{P}{\omega}_{2k} + \overset{R}{\omega}_{2k}$$

follows the pattern given explicitly for the coefficients σ_{2k} earlier.

where we recall relations (8.19b,c).

If we introduce s_1 from (10.21) into (10.45a), the integration of the latter is then a repeated quadrature introducing two arbitrary constants: these are then determined from (10.45b) yielding the unique solution for the principal component of the mean displacement. We omit this calculation and consider the boundary value problem (10.45) solved.

From (10.44), we obtain for the residual component

$$W_R = W_{R0} + xW'_{R0}$$
$$- \frac{2}{35}\frac{h}{E}\left\{ s_3(x) - xs_3'(0) - \frac{1}{18}h^2\left[s_3''(x) - xs_3'''(0) - \frac{1}{11}[s_5''(x) - xs_5'''(0)]\right]\right\} \tag{10.46}$$

which, when added to the solution of (10.45), gives the combined expression for the mean displacement.

We may also decompose the mean rotation in accordance with the prescription (8.21) and (8.22). For the principal and residual components, we have, respectively,

$$\Omega = \frac{(1+\nu)}{E}\frac{1}{3}hs_1' + \frac{1}{2}p_x^* - W', \quad \Omega_R = -W_R' \tag{10.47a,b}$$

the combination of which, on noting (10.43), adds up to formula (10.32a) for the mean rotation.

The remainder of this section is concerned with the analysis of the residual axial stress.

For the system of equations (7.10c) the relevant elements in the sequence $\{\phi_n\}$ are given, in terms of the transverse normal stress coefficients, by the restricted form of (7.7c) with odd index, namely

$$\phi_{2k+1} = h^2\left[\frac{\sigma_{2k-1}''}{(4k-1)(4k+1)} - \frac{2\sigma_{2k+1}''}{(4k+1)(4k+5)} + \frac{\sigma_{2k+3}''}{(4k+5)(4k+7)}\right], \; k \geq 1 \tag{10.48}$$

In this system of relations, the quantity σ_1 appears only in the case $k = 1$, namely in the expression for ϕ_3: writing this case separately and substituting for σ_1 from (10.22), we have the alternate form

$$\phi_3 = h^2[-\frac{2}{45}\sigma_3'' + \frac{1}{99}\sigma_5''] + \frac{1}{525}h^4 s_3'''' + \frac{1}{25}(p + 2hf_0)'' + \frac{1}{150}hp_x^{*\,'''} \tag{10.49a}$$

$$\phi_{2k+1} = h^2 \left[\frac{\sigma''_{2k-1}}{(4k-1)(4k+1)} - \frac{2\sigma''_{2k+1}}{(4k+1)(4k+5)} + \frac{\sigma''_{2k+3}}{(4k+5)(4k+7)} \right], \; k \geq 2 \tag{10.49b}$$

From (10.23) and (10.49), it is now evident that relations (7.7c) with odd index, namely

$$\phi_{2k+1} + 2\sigma_{2k+1} + s_{2k+1} = 0, \; k \geq 1 \tag{10.50}$$

constitute a self-contained system of differential equations for the residual axial stresses to be solved subject to the boundary conditions (10.16).

As with the problem of stretching, we now introduce notation that allows the transformations (10.23) and (10.49) together with the boundary value problem consisting of (10.50) and (10.16) to be written in vector form. We define the infinite dimensional vector functions S_o, Σ_o and Φ_o by setting

$$S_o = \{s_{2k+1}, 1 \leq k \leq \infty\} = \{s_3, s_5, s_7, \ldots, s_{2k+1}, \ldots\} \tag{10.51a}$$

$$\Sigma_o = \{\sigma_{2k+1}, 1 \leq k \leq \infty\} = \{\sigma_3, \sigma_5, \sigma_7, \ldots, \sigma_{2k+1}, \ldots\} \tag{10.51b}$$

$$\Phi_o = \{\phi_{2k+1}, 1 \leq k \leq \infty\} = \{\phi_3, \phi_5, \phi_7, \ldots, \phi_{2k+1}, \ldots\} \tag{10.51c}$$

in each of which the k-th element has suffix $(2k+1)$. For the specified edge quantities, we introduce the vectors $S_o^{\mp}$ and $D_o^{\mp}$ where

$$S_o^{\mp} = \{s_{2k+1}^{\mp}, 1 \leq k \leq \infty\} = \{s_3^{\mp}, s_5^{\mp}, s_7^{\mp}, \ldots, s_{2k+1}^{\mp}, \ldots\} \tag{10.52a}$$

$$D_o^{\mp} = \{d_{2k+1}^{\mp}, 1 \leq k \leq \infty\} = \{d_3^{\mp}, d_5^{\mp}, d_7^{\mp}, \ldots, d_{2k+1}^{\mp}, \ldots\} \tag{10.52b}$$

so that the system of equations (10.50) with the edge conditions (10.16) become

$$\Phi_o + 2\Sigma_o + S_o = 0 \tag{10.53a}$$

$$x = \mp a, \quad S_o = S_o^{\mp}, \quad hS_o' = D_o^{\mp} \tag{10.53b}$$

The system (10.53) is the vector formulation of the boundary value problem for the residual stresses.

In the vector form of the transformation (10.23) and (10.49) the inhomogeneous terms due, respectively, to the applied axial and transverse forces will be written by means of the vectors

$$F_o^{(A)} = \{hp_x^{*\prime}, 0, 0, \ldots\}, \quad F_o^{(T)} = \{(p + 2hf_0), 0, 0, \ldots\} \tag{10.54a,b}$$

in each of which only the first terms is nonzero. The operator for the transformation (10.23), which also occurs in (10.49), will be constructed from the one-sided infinite tridiagonal matrix M_o, where

$$M_o = [m^{(o)}_{k,j}]_1^\infty \tag{10.55}$$

with both suffices ranging from one to infinity. The nonzero elements are given by

$$m^{(o)}_{k,k} = \frac{1}{(4k+1)(4k+5)}, \quad m^{(o)}_{k,k+1} = -\frac{1}{(4k+5)(4k+7)}, \quad 1 \le k \le \infty \tag{10.56a,b}$$

$$m^{(o)}_{k,k-1} = -\frac{1}{(4k-1)(4k+1)}, \quad 2 \le k \le \infty \tag{10.56c}$$

while all other elements vanish, namely

$$m^{(o)}_{k,j} = 0 \qquad \begin{matrix} 1 \le j \le k-2, \; k \ge 3 \\ k+2 \le j \le \infty, \; k \ge 1 \end{matrix} \tag{10.56d}$$

From the above matrix M_o and the basis element matrix E_{11}, we form the matrix functions $L_o(\lambda)$ and $J(\lambda)$, respectively by scalar multiplication by the factor λ: thus

$$L_o(\lambda) = \lambda M_o = [m^{(o)}_{k,j}\lambda]_1^\infty \tag{10.57a}$$

$$J(\lambda) = \lambda E_{11} = [\delta_{k1}\delta_{j1}\lambda]_1^\infty \tag{10.57b}$$

where, in the latter, we have used the Knocker delta notation.

Again, it is convenient to use the scaled coordinates ξ and the mensural ratio α, namely

$$\xi = \frac{x}{h}, \quad \alpha = \frac{a}{h} \tag{10.58a,b,}$$

and we use D to denote the associated derivative element so that

$$D = \frac{d}{d\xi} = h\frac{d}{dx} \tag{10.59}$$

If we now construct the operators $L_o(D^2)$ and $J(D^2)$ by setting

$$L_o(D^2) = [m^{(o)}_{k,j} D^2]_1^\infty, \quad J(D^2) = [\delta_{k1}\delta_{j1}D^2]_1^\infty \tag{10.60a,b}$$

then, in vector form, the transformation (10.23) reads

$$\Sigma_o = -L_o(D^2)S_o - \frac{1}{10}[F_o^{(A)} + F_o^{(T)}] \tag{10.61}$$

while the system (10.49) takes the form

$$\Phi_o = -L_o(D^2)\Sigma_o + \frac{1}{525}[J(D^2)]^2 S_o + \frac{1}{25}D^2[F_o^{(T)} + \frac{1}{6}F_o^{(A)}] \tag{10.62}$$

For the formula giving Φ_o directly in terms of S_o, we introduce Σ_o from (10.61) into (10.62): if we use the fact that

$$L_o(D^2)[F_o^{(A)} + F_o^{(T)}] = \frac{2}{45}D^2[F_o^{(A)} + F_o^{(T)}] \tag{10.63}$$

we obtain

$$\Phi_o = \{[L_o(D^2)]^2 + \frac{1}{525}[J(D^2)]\}S_o + \frac{2}{45}D^2[F_o^{(T)} + \frac{1}{4}F_o^{(A)}] \ . \tag{10.64}$$

The explicit form for the differential equation satisfied by S_o follows from the introduction of Σ_o from (10.61) and Φ_o from (10.64) into (10.53a): we also write the edge conditions (10.53b) in terms of the normalized notation (10.58) so that the statement (10.53) of the boundary value problem assumes the form

$$\{[L_o(D^2)]^2 + \frac{1}{525}[J(D^2)]^2 - 2L_o(D^2) + I\}S_o$$

$$= \frac{1}{5}[F_o^{(T)} + F_o^{(A)}] + \frac{2}{45}D^2[F_o^{(T)} + \frac{1}{4}F_o^{(A)}] \tag{10.65a}$$

$$S_o(\mp\alpha) = S_o^{\mp}, \quad DS_o(\mp\alpha) = D_o^{\mp} \tag{10.65b}$$

in which we again use I to denote the one-sided infinite unit matrix.

As in the problem of stretching, it is appropriate to focus attention on the homogeneous equations: we take

$$p_x^{*\prime} = 0, \quad p = 0, \quad f_0 = 0 \tag{10.66}$$

so that

$$F_o^{(A)} = 0, \quad F_o^{(T)} = 0 \tag{10.67}$$

and the inhomogeneous terms in (10.65a) vanish. Replacing the right hand side of (10.65a) by zero, we have

$$\{[L_0(D^2) - I]^2 + \frac{1}{525}[J(D^2)]^2\}S_0 = 0 \tag{10.68}$$

where we have rearranged the terms in the differential operator.

The subsequent analysis now follows the same pattern as was outlined for the problem of stretching.

Background Survey

The history of beam theory begins with the publication in 1638 of the inquiry of Galileo [12] on the determination of the axis about which a cantilever beam tends to turn under loading. In the absence of a formula relating deformation to force, the question had to remain unanswered until the enunciation of the stress-strain law of Hooke [23] in 1678. Some years later Mariotte [28], who had independently formulated Hooke's Law, applied it to Galileo's problem: assuming that the resistance to flexure arises from the straining of the fibers, he concluded that the axis of flexure lies along the midline of the beam-section. In its application to the problem of the elastica, the method was later extended by James Bernoulli [4], for the determination of the curve assumed by the centerline. From the latter result Euler [10] deduced the moment-curvature relation, which he then took as the basis for his subsequent analysis.

In the later more systematic treatment by Coulomb [7], published in 1776, account is taken of the finite dimension of the beam-section. Considering the resistance to flexure as induced by the extension and contraction of the longitudinal fibers, in accordance with Hooke's Law, Coulomb formulated the hypothesis that the deformation occurs in such a manner that normal sections are displaced into normal sections: he thereby deduced classical beam theory. His analysis preceded, by three decades, the definition of the modulus of elasticity and the introduction of the concept of shear strain by Young [38] in 1807.

The formulation of the general equations of elasticity by Cauchy [5] in 1827 provided the framework for the possible refinement of the classical theory. This motivation led to the celebrated papers of Saint Venant [34] in 1855. This work has already been cited in the General Introduction, where also may be found references to the later developments by other investigators who followed the semi-inverse procedure of Saint Venant in their treatment of beam theory.

Recognizing the potential for obtaining insight on the validity of Saint Venant's Principle, Papkovitch [31], in 1940, formulated and analyzed the problem of the elastic strip within the framework of a two-dimensional theory of elasticity: this was closely followed by the related work of Fadle [11]. A decade later there appeared the further study by Smith [35] and the discussion and extension of Papkovitch's method by Greenberg [13]. There followed the intensive investigations of Horvay [24] and Horvay and Born [25]. The subsequent papers of Benthem [2], Gusein-Zade [16] and Johnson and Little [26], include references to other related literature, while the more recent studies of Gregory [14(a), (b)] and Gregory and Wan [15] attest to the continuing interest in the problem.

We have seen that the method of reduction followed here leads to an infinite system of linear fourth order ordinary differential equations with constant coefficients. Infinite systems first attracted attention following the occurrence of such a system in the Lunar Theory of Hill [20], published in 1877, which led to the subsequent investigations of Poincaré [32] and von Koch [27]. In the years 1904-1910, there appeared the series of monumental papers of Hilbert later collected in book form [19]. Therein the concept of complete continuity is introduced as the requirement appropriate for a discrete spectrum: Hilbert also formulated the conditions on the matrix elements sufficient to guarantee complete continuity for the associated operator. The book by Hellinger and Toeplitz [17] records the results over the next decade including the investigations of spectral theory by Courant and Weyl. The later books of Wintner [37], Stone [36], Cooke [6], Riesz and Nagy [33] and Hille and Phillips [22] treat various aspects of the theory. Of the more recent results on spectral theory we mention the work of Newburgh [30] on the variation of spectra and the analysis of tridiagonal operators by Duren [9].

For a full discussion of the development of the theory, we refer to the excellent historical surveys of Bernkoff [3]. Of the works dealing specifically with systems of ordinary differential equations with infinitely many unknowns, we note the papers of Helly [18], Doob [8], Hille [21(a)], and the more recent investigation by Agmon and Nirenberg [1]. Other features of the theory are treated at greater length in the book by Massera and Schaffer [29], where further references may be found. Certain aspects of such systems are also briefly discussed in the Lectures of Hille [21(b)].

References

1. Agmon, S. and Nirenberg, L., Communications in Pure and Applied Mathematics, vol. 16, 1963.
2. Benthem, J. P., Quarterly Journal of Mechanics and Applied Mathematics, vol. 16, 1963.
3. Bernkoff, M., (a) Archive for the History of the Exact Sciences, vol. 3, 1966.
   (b) Archive for the History Exact Sciences, vol. 4, 1968.
4. Bernoulli, J., *Collected Works,* t. 2, Geneva, 1744.
5. Cauchy, A. L., *Exercises de Mathématiques,* Paris, 1827-28.
6. Cooke, R., *Infinite Matrices and Sequence Spaces,* MacMillan, 1950; New York, Dover, 1960.
7. Coulomb, C. A., Mémoires par divers savants, Paris, 1776.
8. Doob, P. L., Transactions of the American Mathematical Society, vol. 58, 1945.
9. Duren, P. L., Transactions of the American Mathematical Society, vol. 99, 1961.
10. Euler, L., *Methodus Invenienali Lineas Curvas Maximi Minimive Proprietate Gaudentes,* Lausanne, 1744.
11. Fadle, J., Ingenieur Archivium, vol. 11, 1941.
12. Galileo Galilei, *Discorsi e Dimostrazioni Matematische, Leiden,* 1638.
13. Greenberg, G. A., PMM (Applied Mathematics and Mechanics), vol. 17, 1953.
14. Gregory, R. D., (a) Journal of Elasticity, vol. 9, 1979.
    (b) Journal of Elasticity, vol. 10, 1980.

15. Gregory, R. D. and Wan, F. Y. M., Decaying States of Plane Strain in a Semi-infinite strip and boundary conditions for Plate Theory. Inst. Appl. Mech. and Statistics, Univ. B.C., Technical Report 81-17, Sept. 1981.
16. Gusein-Zade, M. I., PMM (Applied Mathematics and Mechanics), vol. 29, 1965.
17. Hellinger, E. and Toeplitz, O., *Integralgleichungen und Gleichungen mit Unendlichvielen Unbekannten,* Leipzig and Berlin, 1928.
18. Helly, E., Monatshefte für Mathematik und Physik, vol. 31, 1921.
19. Hilbert, D., *Grundzuge einer allgemeinen Theorie der Linearen Integralgleichungen,* Leipzig and Berlin, 1912.
20. Hill, G. W., (a) *On the part of the Motion of the lunar perigee which is a function of the motions of the sun and the moon.* Cambridge, Mass., 1877.
(b) Acta Mathematica, vol. 8, 1886.
21. Hille, E., (a) Annali di Matematia Pura et Applicata (4), 55, 1961.
(b) *Lectures on Ordinary Differential Equations,* Addison-Wesley, 1969.
22. Hille, E. and Phillips, R. S., *Functional Analysis and Semi-Groups,* Providence, R.I., 1957 (American Mathematical Society Colloquium Publications vol. 31).
23. Hooke, R., *De Potentia Restitutiva,* London, 1678.
24. Horvay, G., (a) Journal of Applied Mechanics, vol. 20, 1953.
(b) Journal of Applied Mechanics, vol. 24, 1957.
25. Horvay, G. and Born, J. S. (a) Journal of Mathematics and Physics, vol. 33, 1955.
(b) Journal of Applied Mechanics, vol. 24, 1957.
26. Johnson, M. W. and Little, R. W., Quarterly of Applied Mathematics, vol. 22, 1965.
27. von Koch, H., (a) Acta Mathematica, vol. 15, 1891.
(b) Acta Mathematica, vol. 16, 1892-3.
28. Mariotte, E., *Traité de Mouvement des Eaux,* Paris, 1686.
29. Massera, J. L. and Schaffer, J.J., *Linear Differential Equations and Function Spaces,* New York, 1966.
30. Newburgh, J. D., Duke Mathematical Journal, vol. 18, 1951.
31. Papkovitch, P. R., Dokl. Akad. Nauk. U.S.S.R., vol. 27, 1940.
32. Poincaré, H., Bulletin de Societé de Mathématiques France, vol. 14, 1885-6.
33. Riesz, F. and Sz.-Nagy, B., *Functional Analysis,* New York, 1955.
34. de Saint Venant, B., (a) Mémoires des Savants Etrangers, t. 14, 1855.
(b) Journal de Mathématiques (Liouville), ser. 2, t. 1, 1856.
35. Smith, R. C. T., Australian Journal of Scientific Research, vol. 5, 1951.
36. Stone, M. H., *Linear Transformations in Hilbert Space and Their Application to Analysis,* New York, 1932 (American Mathematical Society Colloquium Publications, vol. 15).
37. Wintner, A., *Spektraltheorie der Unendlichen Matrizen,* Leipzig, 1929.
38. Young, T., *A Course of Lectures on Natural Philosophy and the Mechanical Arts,* London, 1807.

"Learning creates difficulties."
Quintilian

Chapter Two

Plate Theory and the Edge Effects

Introduction

In its general form, a plate is a three-dimensional right cylindrical body having a plane of symmetry called the midplane. The bounding surface, in three distinct parts, consists of a cylindrical component known as the edge-surface, normal to the midplane, together with a pair of faces which, symmetrically placed with respect to the midplane, close the ends of the cylinder. We shall refer to the intersection of the edge-surface with the midplane as the edge-curve: for a finite plate this consists of one or more simple closed curves. The distance between the faces, taken along the normal to the midplane at a given point, measures the thickness of the plate at that point.

The plate configuration will be related to a right cylindrical coordinate system erected on the midplane. The reference system then consists of the system of laminar planes parallel to the midplane together with the family of transversals normal thereto. Those elements of the field quantities lying in the laminar planes will be called the laminar components while those associated with the normal to the midplane will be termed the transverse effects.

The body is composed of a material whose mechanical response to the action of applied forces is in accordance with the Cauchy-Green law for a three-dimensional elastic medium. By considering a material for which the transverse direction is mechanically distinguished, we facilitate the detection of the influence of the transverse stresses on the other field quantities — a feature of particular interest in the theory of elastic plates. With this characterization of the material, it remains to define the underlying force field and the designation of the boundary value problem is then completed by the specification of appropriate conditions at each point of the bounding surface. As both the displacement and the mixed boundary value problems can be treated by making appropriate modifications in the procedure developed for the stress boundary value problems, we shall confine our analysis to the latter.

The primary concern of plate theory is the investigation of the effects in the interior due to the conditions applied along the edge-surface. In giving priority to the features of primary interest, we shall consider the stress distribution, applied along the edge-surface, in its most general form, but shall suppress most secondary features of the problem by restricting our consideration to homogeneous plates of constant thickness, subject to a purely normal pressure on the faces, in the absence of a force field. Within the proposed method of reduction, the inclusion of the additional terms due to variation in the plate thickness, could be achieved in the manner already followed in the treatment of beam theory, where the procedure for incorporating the more general form of the forces applied to the faces, as well as the body force effects, has also been indicated.

The stress distribution applied to the edge-surface may be considered a linear combination of two distinct parts: the principal part is characterized by the fact that it contributes to the thickness-integrated stress resultants and stress-couples, while the complementary subsidiary part is a self-equilibrating distribution along any generator of the edge-surface. This resolution, facilitated by the Legendre series representation for the stresses, is then reflected in the interior. The principal part is responsible for the principal effects in the interior, the determination of which is the concern of the principal problems of plate theory: the corresponding residual problems deal with the investigation of the residual effects induced by the subsidiary part of the edge-surface stress distribution.

The plate configuration admits the mutual uncoupling of the problems of stretching and bending, respectively associated with the even and odd parts of the laminar stresses in their dependence on the thickness coordinate. In each problem, when the three-dimensional equations have been integrated with respect to the thickness coordinate by the aid of the Legendre series, we are led to a two-dimensional form of the governing equations in terms of the laminar coordinates, in which the thickness-averaged displacements are the only displacement quantities that appear explicitly. By making an appropriate resolution of these mean displacements into their principal and residual parts, we effect the further uncoupling of the derived form of the governing equations into a set involving the principal effects and a system describing the residual effects. A corresponding separation in the boundary conditions then yields the separate formulation of the principal and residual boundary value problems in both the cases of stretching and of bending.

The residual effects are mainly confined to the narrow boundary layers, whose extent is typically of the order of the plate-thickness. The principal problems are concerned mainly, but not exclusively, with effects that persist at a considerable distance from the edge. In particular, we note that, in the principal bending problem, besides the so-called "interior" effects, there also arises the Reissner "edge" effect due to the resultant transverse shear. From the final formulation it will be evident that the nature of this latter effect is substantially different from the boundary layer phenomena associated with the residual problem.

The objectives of the analysis may now be stated under three main headings:

1. the clarification of the distinction between principal and residual effects in plate theory.
2. the formulation of the principal boundary value problems for both stretching and bending;
3. the formulation of the corresponding residual boundary value problems.

In section One we describe the coordinate system and derive the relevant tensor relations to be used in the sequel. The three-dimensional boundary value problem is formulated in Section Two, followed by a normalized reformulation in Section Three. From the transverse integration of the equilibrium equations, we derive, in Section Four, the Legendre series representations for the stress components: the application of the face conditions then yields the equations satisfied by the leading coefficients. A further consequence of these operations is that all elements in the representations for the transverse stresses are expressed in terms of the coefficients appearing in the series for the laminar stresses. By making the corresponding expansions for the stress distributions applied to the edge-surface, we can then reformulate the edge conditions in terms of the laminar stress coefficients: this is completed in Section Five.

Through the transverse integration of the constitutive equations relating the displacements to the stresses, we obtain, in Section Six, the associated Legendre representations for the displacement components, followed, in Section Seven, by the final two-dimensional formulation of the equations in a form involving the laminar stress coefficients and the mean displacements. Having discussed the uncoupling features of this new formulation in Section Eight, we then proceed to the separate consideration of the problem of stretching and bending. Section Nine is devoted exclusively to the problem of stretching. Having recapitulated the relevant formulae, we then resolve the mean displacements into their principal and residual components and thereby obtain the separated formulations of the principal and residual boundary value problems, followed, in each case, by some further reduction leading to the final form of the system of equations describing each problem. The corresponding analysis and reduction of the systems, describing the principal and residual effects in the problem of bending, are treated in Section Ten.

We conclude with a brief historical review together with references.

1. The Coordinate System

Referred to a general planar coordinate system $[x^\alpha, \alpha = 1,2]$, the position vector to an arbitrary point on the midplane may be written

$$\underset{\sim}{r} = \underset{\sim}{r}\,(x^\alpha) = \underset{\sim}{r}\,(x^1, x^2). \tag{1.1}$$

Using the comma notation for partial differentiation, we have for the associated base vectors $\underset{\sim}{g}_\alpha$,

$$\underset{\sim}{g}_\alpha = \underset{\sim}{r}_{,\alpha} = \frac{\partial \underset{\sim}{r}}{\partial x^\alpha}, \quad \alpha = 1,2 \tag{1.2}$$

from which, by scalar composition, we form the symmetric covariant metric tensor $g_{\alpha\beta}$ of the coordinate system, namely

$$g_{\alpha\beta} = \underset{\sim}{g}_\alpha \cdot \underset{\sim}{g}_\beta, \quad \alpha,\beta = 1,2 \tag{1.3}$$

whose determinant g is clearly given by

$$g = \det[g_{\alpha\beta}]_1^2 = g_{11}\, g_{22} - g_{12}^2 \tag{1.4}$$

The base vectors and metric tensor so defined induce the related dual systems. Following the convention that summation is implied over repeated (dummy) indices, and using the Kronecker symbol to denote the unit matrix, we have the system of equations

$$g_{\alpha\lambda}\, g^{\lambda\beta} = \delta_\alpha^\beta, \quad \alpha,\beta,\lambda, = 1,2 \tag{1.5}$$

for the definition of the components of the contravariant metric tensor $g^{\alpha\beta}$ from which we find

$$g^{11} = \frac{g_{22}}{g}, \quad g^{22} = \frac{g_{11}}{g}, \quad g^{12} = -\frac{g_{12}}{g} \tag{1.6}$$

The conjugate base vectors are then formed by setting

$$\underset{\sim}{g}^\alpha = g^{\alpha\lambda}\, \underset{\sim}{g}_\lambda \tag{1.7}$$

from which it follows that

$$\underset{\sim}{g}_\alpha = g_{\alpha\lambda}\, \underset{\sim}{g}^\lambda, \quad \underset{\sim}{g}_\alpha \underset{\sim}{g}^\beta = \delta_\alpha^\beta \tag{1.8a,b}$$

These reciprocal relations between the two systems of base vectors expedite the mutual conversion of the dual-representations, namely the contravariant and covariant forms: such transformations are then reflected in the lowering or raising of the tensor indices as the context requires.

With the metric $g_{\alpha\beta}$ there is associated the two dimensional covariant permutation tensor $\epsilon_{\alpha\beta}$ with components

$$\epsilon_{11} = \epsilon_{22} = 0, \quad \epsilon_{12} = -\epsilon_{21} = \sqrt{g} \tag{1.9}$$

where g is given by (1.4). The corresponding contravariant tensor is defined by

$$\epsilon^{\alpha\beta} = g\epsilon^{\alpha\lambda} g\epsilon^{\beta\mu} \epsilon_{\lambda\mu} \tag{1.10}$$

which, in explicit form, gives

$$\epsilon^{11} = \epsilon^{22} = 0\,, \quad \epsilon^{12} = -\epsilon^{21} = \frac{1}{\sqrt{g}} \tag{1.11}$$

By means of the permutation tensor, relations (1.6) may be written

$$g^{\alpha\beta} = \epsilon^{\alpha\lambda} \epsilon^{\beta\mu} g_{\lambda\mu} \tag{1.12}$$

as can be readily verified by introducing the components from (1.11).

For the two dimensional space with metric $g_{\alpha\beta}$, the Christoffel symbols of the first and second kinds are respectively written

$$[\alpha,\beta:\mu] = \frac{1}{2}\,[g_{\alpha\mu,\beta} + g_{\beta\mu,\alpha} - g_{\alpha\beta,\mu}] \tag{1.13a}$$

$$\left\{ \begin{matrix} \lambda \\ \alpha\,\beta \end{matrix} \right\} = g^{\lambda\mu}[\alpha,\beta:\mu] = \frac{1}{2}\, g^{\lambda\mu}[g_{\alpha\mu,\beta} + g_{\beta\mu,\alpha} - g_{\alpha\beta,\mu}] \tag{1.13b}$$

The occurrence of the latter symbol in the formula for the covariant derivative will follow from the fact that for the base vectors we have

$$\underset{\sim}{g}_{\alpha,\beta} = \left\{ \begin{matrix} \lambda \\ \alpha\,\beta \end{matrix} \right\} \underset{\sim}{g}_{\lambda}\,, \quad \underset{\sim}{g}^{\alpha}_{,\beta} = -\left\{ \begin{matrix} \alpha \\ \lambda\,\beta \end{matrix} \right\} \underset{\sim}{g}^{\lambda} \tag{1.14a,b}$$

The former can be immediately verified by partial differentiation of formulae (1.3) and forming appropriate combinations: the latter then follows as a corollary from differentiation of relation (1.8b). Of particular importance will be the trace of formula (1.13b) namely

$$\left\{ \begin{matrix} \lambda \\ \alpha\,\lambda \end{matrix} \right\} = \frac{1}{2}\, g^{\lambda\mu} g_{\lambda\mu,\alpha} \tag{1.15}$$

If, in the matrix $[g_{\alpha\beta}]$, we denote the cofactor of the element $g_{\alpha\beta}$ by $g_*^{\alpha\beta}$ so that, noting (1.6), we have

$$g_*^{\alpha\beta} = g\, g^{\alpha\beta} \tag{1.16}$$

then for relation (1.15), we have the alternate form

$$\left\{ \begin{matrix} \lambda \\ \alpha\,\lambda \end{matrix} \right\} = \frac{1}{2g}\, g_{*}^{\lambda\mu}\, g_{\lambda\mu,\alpha} \tag{1.17}$$

Applying the rule for the differentiation of a determinant, we see that formula (1.17) is equivalent to

$$\left\{ \begin{matrix} \lambda \\ \alpha\,\lambda \end{matrix} \right\} = \frac{1}{2g}\,\frac{\partial g}{\partial x^{\alpha}} \tag{1.18}$$

giving the important relation

$$\frac{\partial}{\partial x^{\alpha}} \sqrt{g} = \sqrt{g} \left\{ \begin{matrix} \lambda \\ \alpha\,\lambda \end{matrix} \right\} \tag{1.19}$$

which we shall use in our derivation of the equilibrium equations.

In a planar vector field, referred either to the base vectors $\underset{\sim}{g}_{\alpha}$ or to the conjugate system $\underset{\sim}{g}^{\alpha}$, the operation of partial differentiation with respect to one of the coordinate variables yields a second order tensor, whose components are the covariant derivatives of the components of the original vector. Thus in the case of the vector

$$\underset{\sim}{a} = a^{\lambda}\, \underset{\sim}{g}_{\lambda} = a_{\lambda}\, \underset{\sim}{g}^{\lambda} \tag{1.20}$$

we have for the partial derivative

$$\underset{\sim}{a}_{,\alpha} = [a^{\lambda}{}_{,\alpha} + \left\{ \begin{matrix} \lambda \\ \mu\,\alpha \end{matrix} \right\} a^{\mu}]\, \underset{\sim}{g}_{\lambda} \tag{1.21a}$$

$$= [a_{\lambda,\alpha} - \left\{ \begin{matrix} \mu \\ \lambda\,\alpha \end{matrix} \right\} a_{\mu}]\, \underset{\sim}{g}^{\lambda} \tag{1.21b}$$

We shall use a single vertical stroke to denote covariant differentiation in the two dimensional space with metric $g_{\alpha\beta}$: hence, for the contravariant components of the vector, we have as defining formula for the covariant derivative

$$a^{\lambda}|_{\alpha} = a^{\lambda}{}_{,\alpha} + \left\{ \begin{matrix} \lambda \\ \mu\,\alpha \end{matrix} \right\} a^{\mu} \tag{1.22a}$$

while for the covariant components the corresponding relations read

$$a_\lambda|_\alpha = a_{\lambda,\alpha} - \left\{ \begin{matrix} \mu \\ \lambda\alpha \end{matrix} \right\} a_\mu \tag{1.22b}$$

and expressions (1.21) for the partial derivative of a vector take the compact form

$$\underset{\sim}{a}_{,\alpha} = a^{\lambda}|_\alpha \underset{\sim}{g}_\lambda = a_\lambda|_\alpha \underset{\sim}{g}^{\lambda} \tag{1.23}$$

For the components of tensors of higher order, the definition of the covariant derivative is a direct extension of the formula for that of the components of a vector: thus, for a second order tensor $a^{\alpha\beta}$ contravariant in both indices and defined in the space with metric $g_{\alpha\beta}$, the covariant derivative has the form

$$a^{\alpha\beta}|_\lambda = a^{\alpha\beta}{}_{,\lambda} + \left\{ \begin{matrix} \alpha \\ \mu\lambda \end{matrix} \right\} a^{\mu\beta} + \left\{ \begin{matrix} \beta \\ \mu\lambda \end{matrix} \right\} a^{\alpha\mu} \tag{1.24}$$

generalizing relation (1.22a) for the covariant derivative of tensors of first order.

In the three dimensional Euclidean space, we first consider the general coordinate system $[x^i, i=1,2,3]$ and write for the position vector to an arbitrary point

$$\underset{\sim}{R} = \underset{\sim}{R}(x^i) = \underset{\sim}{R}(x^1, x^2, x^3) \tag{1.25}$$

Associated with the coordinate system $[x^i, i=1,2,3]$, we have the base vectors

$$\underset{\sim}{G}_i = \underset{\sim}{R}_{,i} = \frac{\partial \underset{\sim}{R}}{\partial x^i}, \quad i = 1,2,3 \tag{1.26}$$

in terms of which we define the covariant metric tensor G_{ij} by setting

$$G_{ij} = \underset{\sim}{G}_i \underset{\sim}{G}_j, \quad i,j = 1,2,3 \tag{1.27}$$

with determinant

$$G = \det[G_{ij}]_1^3 \tag{1.28}$$

The components of the contravariant metric tensor G^{ij} constitute the inverse matrix determined by

$$G_{ik}\, G^{kj} = \delta_i^j \tag{1.29}$$

where the Kronecker delta on the right denotes the unit matrix: the relations

$$\underset{\sim}{G}^{i} = G^{ik}\,\underset{\sim}{G}_{k} \tag{1.30}$$

then give the conjugate system of base vectors, and we further note that

$$\underset{\sim}{G}_{i} = G_{ik}\,\underset{\sim}{G}^{k}\,,\quad \underset{\sim}{G}_{i}\cdot\underset{\sim}{G}^{j} = \delta_{i}^{j} \tag{1.31}$$

Using the letter Γ to distinguish the symbols in the three dimensional space, we write

$$\Gamma_{jk:i} = \frac{1}{2}[G_{ij,k} + G_{ik,j} - G_{jk,i}] \tag{1.32a}$$

$$\Gamma^{i}_{jk} = G^{il}\,\Gamma_{jk:l} = \frac{1}{2}\,G^{il}\,[G_{lj,k} + G_{lk,j} - G_{jk,l}] \tag{1.32b}$$

respectively for the Christoffel symbols of the first and second kinds associated with the metric G_{ij}, and, as in the two dimensional case, it can be easily checked that, for the derivatives of the base vectors, we have

$$\underset{\sim}{G}_{j,k} = \Gamma^{i}_{jk}\,G_{i} \tag{1.33}$$

Hence, for an arbitrary vector field $\underset{\sim}{A}$, for which we have the component representations

$$\underset{\sim}{A} = a^{i}\,\underset{\sim}{G}_{i} = a_{i}\,\underset{\sim}{G}^{i} \tag{1.34}$$

partial differentiation with respect to the coordinate variables yields

$$\underset{\sim}{A}_{,j} = a^{i}\|_{j}\,\underset{\sim}{G}_{i} = a_{i\|j}\,\underset{\sim}{G}^{i} \tag{1.35}$$

where we have used a double vertical stroke to distinguish covariant differentiation in the three dimensional space with metric G_{ij}: that is, we have set

$$a^{i}\|_{j} = a^{i}_{,j} + \Gamma^{i}_{jk}\,a^{k},\quad a_{i\|j} = a_{i,j} - \Gamma^{k}_{ij}\,a_{k} \tag{1.36a,b}$$

while for a second order tensor with components a^{ij}, contravariant in both indices, we would write

$$a^{ij}\|_{k} = a^{ij}_{,k} + \Gamma^{i}_{lk}\,a^{lj} + \Gamma^{j}_{lk}\,a^{il} \tag{1.37}$$

as the defining formula for the covariant derivative.

For a plate configuration, the natural frame of reference is the general cylindrical coordinate system consisting of the planar system $[x^{\alpha}, \alpha = 1,2]$ augmented by the transverse coordinate z measured normal to the midplane. We, therefore, specialize the general system $[\mathrm{x}^{i}, i = 1,2,3]$ by identifying the first two coordinates with those of the planar system $[x^{\alpha}, \alpha = 1,2]$ and equating the third with the transverse coordinate z: explicitly, we have

$$\mathrm{x}^{\alpha} = x^{\alpha}, \ \alpha = 1,2, \ \mathrm{x}^{3} = z \tag{1.38}$$

so that, in place of (1.25), we write for the position vector to an arbitrary point

$$\underset{\sim}{R} = \underset{\sim}{R}(\mathrm{x}^{i}) = \underset{\sim}{R}(x^{\alpha}, z) \tag{1.39}$$

which may now be considered the resultant of adding the transverse segment to the position vector of the corresponding point on the midplane: noting that, for the latter point, we have

$$\underset{\sim}{r}(x^{\alpha}) = \underset{\sim}{R}_{0} = \underset{\sim}{R}(x^{\alpha}, 0) \tag{1.40}$$

and letting $\underset{\sim}{n}$ denote the constant unit vector normal to the midplane, we see that the position vector (1.39) has the resolved form

$$\underset{\sim}{R} = \underset{\sim}{R}(\mathrm{x}^{i}) = \underset{\sim}{R}(x^{\alpha}, z) = \underset{\sim}{r}(x^{\alpha}) + z\underset{\sim}{n} \tag{1.41}$$

which enables us to relate the elements of the spatial reference system to those of the associated planar system.

With the defining relations (1.26), it follows from (1.41) and (1.2) that for the cylindrical coordinate system, the base vectors are

$$\underset{\sim}{G}_{\alpha} = \underset{\sim}{g}_{\alpha}, \ \alpha = 1,2; \ \underset{\sim}{G}_{3} = \underset{\sim}{n} \tag{1.42a,b}$$

and hence, the components of the metric tensor G_{ij} are given by

$$G_{\alpha\beta} = g_{\alpha\beta}, \ \alpha, \beta = 1,2, \tag{1.43a}$$

$$G_{\alpha 3} = 0, \ \alpha = 1,2: \ G_{33} = 1 \tag{1.43b,c}$$

so that the determinant of the metric G_{ij} is identical with that of the metric $g_{\alpha\beta}$, namely

$$G = g \tag{1.44}$$

Using relations (1.43) and (1.44) in our consideration of the defining equations (1.29) and (1.30), it is evident that the contravariant components of the spatial metric tensor are reducible to those of the planar metric tensor, namely

$$G^{\alpha\beta} = g^{\alpha\beta}, \quad \alpha,\beta = 1,2 \tag{1.45a}$$

$$G^{\alpha 3} = 0 \quad , \alpha = 1,2: \; G^{33} = 1 \tag{1.45b,c}$$

while for the conjugate base vectors, we have

$$\underset{\sim}{G}^{\alpha} = \underset{\sim}{g}^{\alpha}, \; \alpha = 1,2: \; \underset{\sim}{G}^{3} = n \tag{1.46a:b}$$

Applying relations (1.43) and (1.45) to formulae (1.32) for the Christoffel symbols, we see that those symbols with index three in any position vanish: in particular, for the Christoffel symbols of the second kind, we have

$$\Gamma^{3}_{jk} = 0, \; \Gamma^{i}_{3k} = 0, \; \Gamma^{i}_{j3} = 0, \; i,j,k = 1,2,3 \tag{1.47}$$

while the nonvanishing elements coincide with the corresponding symbols associated with the two-dimensional metric $g_{\alpha\beta}$ as given in (1.13b), namely

$$\Gamma^{\lambda}_{\alpha\beta} = \left\{ \begin{matrix} \lambda \\ \alpha\;\beta \end{matrix} \right\}, \; \lambda,\alpha,\beta = 1,2, \tag{1.48}$$

From these relations, we see that, in the cylindrical coordinate system $[x^{\alpha}, z; \alpha = 1,2]$, covariant differentiation with respect to either of the laminar coordinates x^{α} is unaffected by the presence of the transverse dimension, while for the transverse coordinate covariant differentiation is identical with partial differentiation. In particular, for the vector field $\underset{\sim}{A}$ of (1.34), whose covariant representation may now be written

$$\underset{\sim}{A} = a_{\alpha}\, \underset{\sim}{g}^{\alpha} + a_3\, \underset{\sim}{n} \tag{1.49}$$

we have, for the covariant derivatives of the normal and laminar components respectively

$$a_{3\|3} = a_{3,z}, \; a_{3\|\alpha} = a_{3,\alpha} \tag{1.50a,b}$$

$$a_{\alpha\|3} = a_{\alpha,3} \quad a_{\alpha\|\beta} = a_{\alpha|\beta} \tag{1.50c,d}$$

which, as we shall see, will result in a considerable simplication in our expressions for the strain measures.

For the description of the plate we first note that referred to the planar coordinate system $[x^{\alpha}, \alpha = 1,2]$, the equation of the edge curve, along which the bounding surface intersects the midplane, may be written in the parametric form

$$\mathcal{C}: \qquad x^{\alpha} = \overline{x^{\alpha}(\eta)}, \ \alpha = 1,2, \ 0 \leq \eta \leq 1 \ . \qquad (1.51)$$

where the $\overline{x^{\alpha}}$ are given functions of the appropriately chosen normalized running parameter η on the curve. The nonparametric form

$$\mathcal{C}: \qquad Y(x^{\alpha}) = Y(x^1, x^2) = 0 \qquad (1.52)$$

follows from the elimination of η from (1.51) and hence, the connected set

$$\mathcal{A}: \qquad Y(x^{\alpha}) \leq 0 \qquad (1.53)$$

defines the region occupied by the projection of the plate on the midplane. Accordingly, if we denote the half-thickness of the plate by h, then, in cylindrical coordinates, the boundary consists of the cylindrical edge surface

$$\mathcal{B}: \qquad x^{\alpha} = \overline{x^{\alpha}(\eta)}, \ \alpha = 1,2, \ 0 \leq \eta \leq 1; \ -h \leq z \leq h \qquad (1.54a)$$

together with the pair of faces

$$z = \mp h, \ Y(x^{\alpha}) \leq 0 \qquad (1.54b)$$

and hence, the region

$$\mathcal{R}: \qquad Y(x^{\alpha}) < 0, \ |z| < h \qquad (1.55)$$

defines the interior of the plate.

We shall use a bar to indicate when quantities are evaluated on the edge so that on the curve (1.51) we write for the base vectors

$$\overline{\underset{\sim}{g}_{\alpha}} = \overline{\underset{\sim}{g}_{\alpha}(\eta)} = \left[\underset{\sim}{g}_{\alpha}(x^{\lambda})\right]_{x^{\lambda} = \overline{x^{\lambda}(\eta)}} \qquad (1.56a)$$

$$\overline{\underset{\sim}{g}^{\alpha}} = \overline{\underset{\sim}{g}^{\alpha}(\eta)} = \left[\underset{\sim}{g}^{\alpha}(x^{\lambda})\right]_{x^{\lambda} = \overline{x^{\lambda}(\eta)}} \qquad (1.56b)$$

and for the metric and permutation tensors

$$\overline{g_{\alpha\beta}} = \overline{g_{\alpha\beta}(\eta)} = \left[g_{\alpha\beta}(x^{\lambda})\right]_{x^{\lambda}=\overline{x^{\lambda}(\eta)}} \tag{1.57a}$$

$$\overline{\epsilon_{\alpha\beta}} = \overline{\epsilon_{\alpha\beta}(\eta)} = \left[\epsilon_{\alpha\beta}(g)\right]_{g=\overline{g(\eta)}} \tag{1.57b}$$

where in the latter we have noted (1.9) and set

$$\overline{g(\eta)} = \det[\overline{g_{\alpha\beta}(\eta)}]_1^2 \tag{1.58}$$

For the curve (1.51), the contravariant components of the tangent vector are

$$\chi^{\alpha}(\eta) = \frac{d}{d\eta}\overline{x^{\alpha}(\eta)} \tag{1.59}$$

and hence, the covariant quantities

$$y_{\alpha}(\eta) = \overline{\epsilon_{\alpha\lambda}(\eta)}\,\chi^{\lambda}(\eta) \tag{1.60}$$

constitute the corresponding normal vector. Defining the associated normalized components by

$$\mathfrak{n}_{\alpha}(\eta) = \frac{y_{\alpha}(\eta)}{\sqrt{y_{\lambda}(\eta)\,y^{\lambda}(\eta)}} \tag{1.61}$$

we have the resolution

$$\underset{\sim}{\mathfrak{n}} = \mathfrak{n}_{\alpha}\,\underset{\sim}{g}^{\alpha} \tag{1.62}$$

for the unit vector $\underset{\sim}{\mathfrak{n}}$ normal to the edge curve.

Since the part of the boundary defined by (1.54a) is a right cylinder erected on the edge curve (1.51), the unit normal at an arbitrary point (η, z) of the edge surface is independent of z and is given by formula (1.62) evaluated at the corresponding point of projection $(\eta, 0)$ on the edge curve. On the remainder of the boundary consisting of the lower and upper faces defined in (1.54b), the unit normal is constant on each face and coincides with the unit transverse vector $\underset{\sim}{\eta}$, except for a change of sign on the lower face only.

As in the above, so in the subsequent notation, Greek indices will consistently have the range (1,2), while Latin indices will range over (1,2,3).

2. The Boundary Value Problem

In a three-dimensional medium referred to a general coordinate system $[x^i, i=1,2,3]$, in which the position vector is denoted by $\underset{\sim}{R}(x^i)$ as in (1.25), and the associated base vectors $\underset{\sim}{G}_i$ and metric tensor G_{ij} are given by (1.26) and (1.27) respectively, if we represent the stress tensor by τ^{ij} and write f^j for the contravariant components of the body force density, then the equilibrium of a material element requires* both the symmetry of the stress tensor

$$\tau^{ij} = \tau^{ji}, \quad i,j = 1,2,3 \tag{2.1}$$

and the satisfaction of the vector equation

$$\frac{\partial}{\partial x^i}[\sqrt{G}\,\tau^{ij}\,\underset{\sim}{G}_j] + \sqrt{G}\,f^j\,\underset{\sim}{G}_j = 0 \tag{2.2}$$

wherein G is the determinant of the metric tensor as given by (1.28), and we again follow the convention of summation over repeated indices. The latter equation is equivalent to the system of tensor equations

$$\tau^{ij}\|_i + f^j = 0 \tag{2.3}$$

where, as previously stated, the double vertical stroke indicates covariant differentiation defined for the metric G_{ij} by (1.37) and (1.32). These equations may now be specialized for the cylindrical coordinate system natural to the plate configuration in accordance with relations (1.41) to (1.50).

An alternative – and possibly more direct – procedure would introduce the specialization directly into equation (2.2).** Giving explicit preference to the third coordinate and omitting the body force terms, equation (2.2) reads

$$\frac{\partial}{\partial x^\alpha}[\sqrt{G}\,(\tau^{\alpha\beta}\,\underset{\sim}{G}_\beta + \tau^{\alpha 3}\,\underset{\sim}{G}_3)] + \frac{\partial}{\partial x^3}[\sqrt{G}\,(\tau^{3\beta}\,\underset{\sim}{G}_\beta + \tau^{33}\,\underset{\sim}{G}_3)] = 0 \tag{2.4}$$

where, as before the Greek indices have the range (1,2). If we now identify the first two coordinates $(x^\alpha, \alpha=1,2)$ with the laminar coordinates $[x^\alpha, \alpha=1,2]$ and equate x^3 with the transverse coordinate z, then recalling relations (1.42) and (1.44) we see that, for the cylindrical coordinate system, equation (2.4) becomes

$$\frac{\partial}{\partial x^\alpha}[\sqrt{g}\,(\tau^{\alpha\beta}\,\underset{\sim}{g}_\beta + \tau^{\alpha 3}\,\underset{\sim}{n})] + \frac{\partial}{\partial z}[\sqrt{g}\,(\tau^{3\beta}\,\underset{\sim}{g}_\beta + \tau^{33}\,\underset{\sim}{n})] = 0 \tag{2.5}$$

* For the derivation of these and other basic equations of the general theory of elasticity use refer to the standard treatise, e.g. "Theoretical Elasticity" by A.E. Green and W. Zerma. Oxford 1954.

** Here the procedure is followed primarily for consistency with our later analysis of equilibrium in a shell where the advantages are more obvious. Actually from our observations in (1.47) and (1.48), it is obvious that equations (2.3) specialize to the system (2.7) in the cylindrical coordinate system.

Performing the differentiation we utilize formulae (1.14a) and (1.19) and rearrange: omitting the common factor $\sqrt{g}$ we obtain

$$[\tau^{\alpha\beta}{}_{,\alpha} + \left\{\begin{matrix}\beta\\ \lambda\,\alpha\end{matrix}\right\} \tau^{\alpha\lambda} + \left\{\begin{matrix}\lambda\\ \alpha\,\lambda\end{matrix}\right\} \tau^{\alpha\beta} + \tau^{3\beta}{}_{,z}]\, \underset{\sim}{g}_\beta$$

$$+ [\tau^{\alpha 3}{}_{,\alpha} + \left\{\begin{matrix}\lambda\\ \alpha\,\lambda\end{matrix}\right\} \tau^{\alpha 3} + \tau^{33}{}_{,z}]\, \underset{\sim}{n} = 0 \tag{2.6}$$

We may now use the notation of covariant differentiation for the coordinate system $[x^\alpha, \alpha = 1,2]$ with metric $g_{\alpha\beta}$ as defined in relations (1.22a) and (1.24) so that the vector equation (2.6) written in component form becomes the system of tensor equations

$$\tau^{\alpha\beta}|_\alpha + \tau^{3\beta}{}_{,z} = 0 \tag{2.7a}$$

$$\tau^{\alpha 3}|_\alpha + \tau^{33}{}_{,z} = 0 \tag{2.7b}$$

If we also give preference to the transverse coordinate in the statement of the symmetric character of the stress tensor, then in place of (2.1) we write

$$\tau^{\alpha\beta} = \tau^{\beta\alpha}, \quad \tau^{\alpha 3} = \tau^{3\alpha} \tag{2.8a,b}$$

as the conditions supplementing equations (2.7) for the equilibrium requirements formulated in the cylindrical coordinate system.

When the displacement vector field $\underset{\sim}{V}$ is referred to the base vectors $\underset{\sim}{G}^i$ giving the resolution

$$\underset{\sim}{V} = v_i\, \underset{\sim}{G}^i \tag{2.9}$$

the linearized strain measures constituting the symmetric linear strain tensor γ_{ij} are defined in terms of the displacement components by

$$\gamma_{ij} = \frac{1}{2}(v_{i\|j} + v_{j\|i}), \quad i,j = 1,2,3. \tag{2.10}$$

in which the covariant differentiation indicated by the double vertical stroke is in accordance with (1.36) and (1.32). For the cylindrical coordinate system, the resolution (2.9) becomes

$$\underset{\sim}{V} = v_\alpha\, \underset{\sim}{g}^\alpha + v_3\, \underset{\sim}{n}$$

and relations (1.50) now give the defining formulae (2.10) a simpler form: the extensional and shear strains associated with the transverse direction are given respectively by

$$\gamma_{33} = v_{3,z} \tag{2.12a}$$

and

$$\gamma_{\alpha 3} = \frac{1}{2}\,(v_{\alpha,z} + v_{3,\alpha})\ ,\ \alpha = 1,2 \tag{2.12b}$$

while for the laminar strains we have

$$\gamma_{\alpha\beta} = \frac{1}{2}\,(v_{\alpha\,|\beta} + v_{\beta\,|\alpha}),\ \alpha,\beta = 1,2 \tag{2.12c}$$

in which the covariant differentiation, indicated by the single vertical stroke, is now in accordance with (1.22b).

For a homogeneous material with linear stress-strain law, the Cauchy-Green constitutive relations have the form

$$\gamma_{ij} = E_{ijkl}\,\tau^{kl} \tag{2.13}$$

in which the elastic parameters E_{ijkl} are constant. Since the coefficient tensor E_{ijkl} shares the symmetry properties of both the stress and strain tensors there can be at most thirty-six independent elements: the existence of a strain energy function implies a further symmetry condition leaving twenty-one independent constants in the coefficient tensor. If we impose the further restriction that the plate material be laminarly isotropic, that is, we require invariance in the mechanical characteristics under changes of direction in laminae parallel to the midplane but admit a distinguished response for the transverse direction, then, the number of independent elastic parameters reduces to five and, when related to the cylindrical coordinate system, the constitutive relations (2.13) take the special form

$$\gamma_{33} = \frac{1}{E_z}\,\tau_{33} - \frac{\nu_*}{E}\,\tau^{\lambda}_{\lambda} \tag{2.14a}$$

$$\gamma_{\alpha 3} = \frac{1}{B}\,\tau_{\alpha 3} \tag{2.14b}$$

$$\gamma_{\alpha\beta} = \frac{(1+\nu)}{E}\,\tau_{\alpha\beta} - g_{\alpha\beta}\cdot\frac{1}{E}\,[\nu\tau^{\lambda}_{\lambda} + \nu_*\,\tau_{33}] \tag{2.14c}$$

in which the covariant components of the stress tensor appearing on the right hand side are in accordance with metric relations (1.43), namely

$$\tau_{\alpha\beta} = g_{\alpha\lambda}\, g_{\beta\mu}\, \overset{\lambda\mu}{\tau}, \quad \tau_{33} = \overset{3}{\tau_3} = \overset{33}{\tau} \tag{2.15a,b}$$

$$\tau_{\alpha 3} = g_{\alpha\lambda}\, \overset{\lambda}{\tau_3} = g_{\alpha\lambda}\, \overset{\lambda 3}{\tau} \tag{2.15c}$$

while the invariant $\overset{\lambda}{\tau_\lambda}$ is the trace of the laminar part of the stress tensor defined by

$$\overset{\lambda}{\tau_\lambda} = g_{\lambda\mu}\, \overset{\lambda\mu}{\tau} \tag{2.15d}$$

The moduli E_z and B reflect the transverse extensibility and transverse shear deformability respectively while the dimensionless ratio ν_* measure the effect of the transverse stress on the laminar strains: the coefficients E and ν are respectively, the Young's modulus and Poisson's ratio associated with the laminar effects. We note that a transversely inextensible material is characterized by the conditions

$$\frac{1}{E_z} = 0, \quad \nu_* = 0 \tag{2.16}$$

while the independent property of transverse shear rigidity is expressed by

$$\frac{1}{B} = 0 \tag{2.17}$$

In the case of complete isotropy, we would have

$$E_z = E, \quad \nu_* = \nu, \quad B = \frac{E}{(1+\nu)} \tag{2.18}$$

the number of independent coefficients being then reduced to two.

Besides the strain components (2.12) there are also associated with the displacement quantities the three components of infinitesimal rotation. In the cylindrical coordinate system, the laminar rotation ψ is defined in terms of the laminar displacements by

$$\psi = \frac{1}{2}(v_{1|2} - v_{2|1}) = \frac{1}{2}(v_{1,2} - v_{2,1}) \tag{2.19}$$

while the transverse components of infinitesimal rotation ω_α are given by

$$\omega_\alpha = \frac{1}{2}(v_{\alpha,z} - v_{3,\alpha}) \tag{2.20}$$

The latter may be put in the alternate form

$$\omega_\alpha = \frac{1}{B}\tau_{\alpha\beta} - v_{3,\alpha} \tag{2.20*}$$

if we first use relation (2.12b) to eliminate $v_{\alpha,z}$ and then substitute for $\gamma_{\alpha 3}$ from the constitutive relation (2.14b).

On the boundary we consider each face subject to a purely normal pressure while on the cylindrical edge we admit the application of a general stress vector field distributed in an arbitrary manner over the surface. If we let p^- and p^+ denote the pressures on the lower and upper faces respectively, then recalling the description (1.54b) for these planar segments we have for the conditions on the lower face

$$z = -h,\ Y(\overset{\lambda}{x}) \le 0:\ \tau^{3\alpha}(\overset{\lambda}{x}, -h) = 0,\ \tau^{33}(\overset{\lambda}{x}, -h) = p^-(\overset{\lambda}{x}) \tag{2.21a,b}$$

while the corresponding form for the conditions on the upper face reads

$$z = +h,\ Y(\overset{\lambda}{x}) \le 0:\ \tau^{3\alpha}(\overset{\lambda}{x}, +h) = 0,\ \tau^{33}(\overset{\lambda}{x}, +h) = p^+(\overset{\lambda}{x}) \tag{2.22a,b}$$

At an arbitrary point (η, z) of the cylindrical surface (1.54a), we let $[Z^i(\eta, z),\ i = 1,2,3]$ represent the components of the applied stress vector, so that the boundary conditions on the edge take the form

$$-h \le z \le \pm h,\ \overset{\lambda}{x} = \overline{\overset{\lambda}{x}(\eta)}: \begin{cases} \overline{\tau^{\alpha\beta}}(\eta, z)\ \mathfrak{n}_\alpha(\eta) = Z^\beta(\eta, z) & \text{(2.23a)} \\[2ex] \overline{\tau^{\alpha 3}}(\eta, z)\ \mathfrak{n}_\alpha(\eta) = Z^3(\eta, z) & \text{(2.23b)} \end{cases}$$

in which the $\mathfrak{n}_\alpha$ are the components of the unit normal vector (1.62) and where we have used the notation

$$\overline{\tau^{\alpha\beta}}(\eta, z) = \left[\tau^{\alpha\beta}(\overset{\lambda}{x}, z)\right]_{\overset{\lambda}{x} = \overline{\overset{\lambda}{x}(\eta)}},\quad \overline{\tau^{\alpha 3}}(\eta, z) = \left[\tau^{\alpha 3}(\overset{\lambda}{x}, z)\right]_{\overset{\lambda}{x} = \overline{\overset{\lambda}{x}(\eta)}} \tag{2.24a,b}$$

to signify the edge values of the relevant stress components.

The combination of relations (2.12) with (2.14) gives the system of stress-displacement relations complementing the equilibrium conditions (2.7) and (2.8): the determination of the solution to this combined system of partial differential equations valid in the region $\mathcal{R}$ defined in (1.55), and satisfying the boundary conditions (2.21), (2.22) and (2.23) constitutes the stress boundary value problem of plate theory.

3. The Normalized Formulation

To obtain a normalized formulation we make a coordinate transformation that leaves the laminar coordinates unchanged and replaces the transverse coordinate by its normalized dimensionless transform. Introducing the thickness variable t by setting

$$t = \frac{z}{h} \tag{3.1}$$

then, since h is constant, we have

$$\frac{\partial t}{\partial x^{\alpha}} = 0, \quad \frac{\partial t}{\partial z} = \frac{1}{h} \tag{3.2a,b}$$

so that only the z-differentiation is affected by the transformation. The plate interior $\mathcal{R}$ defined in (1.55) corresponds in the transformed coordinate system (x^{α}, t) to the region.

$$\mathcal{R}_0: \qquad -1 \le t \le +1, \quad Y(x^{\alpha}) \le 0 \tag{3.3}$$

whose boundary is now described by the pair of faces

$$t = \mp 1, \quad Y(x^{\alpha}) \le 0 \tag{3.4}$$

together with the edge surface

$$-1 \le t \le +1, \; x^{\alpha} = \overline{x^{\alpha}(\eta)} \tag{3.5}$$

In the region $\mathcal{R}_0$ all stress and displacement quantities are now to be considered functions of the x^{α} and t and we accordingly introduce notation reflecting the effect of the transformation (3.1). Starting with the stress tensor, we write for the laminar components

$$\sigma^{\alpha\beta}(x^{\lambda}, t) = \tau^{\alpha\beta}(x^{\lambda}, z) \tag{3.6a}$$

and, while for the transverse normal stress we set

$$\sigma(x^{\lambda}, t) = \tau^{33}(x^{\lambda}, z) \tag{3.6b}$$

the transformed form of the transverse shear components are distinguished by

$$\tau^{\alpha}(x^{\lambda}, t) = \tau^{\alpha 3}(x^{\lambda}, z) = \tau^{3\alpha}(x^{\lambda}, z) \tag{3.6c}$$

In the transformed displacement quantities, we separate the transverse component from the laminar components by setting

$$u_\alpha(x^\lambda,t) = v_\alpha(x^\lambda,z), \quad \alpha = 1,2: \quad w(x^\lambda,t) = v_3(x^\lambda,z) \tag{3.7a:b}$$

The relevant edge values become

$$\overline{\sigma^{\alpha\beta}(\eta,t)} = \left[\sigma^{\alpha\beta}(x^\lambda,t)\right]_{\overline{x^\lambda = x^\lambda(\eta)}} = \overline{\tau^{\alpha\beta}(\eta,z)} \tag{3.8a}$$

$$\overline{\tau^{\alpha}(\eta,t)} = \left[\tau^{\alpha}(x^\lambda,t)\right]_{\overline{x^\lambda = x^\lambda(\eta)}} = \overline{\tau^{\alpha 3}(\eta,z)} \tag{3.8b}$$

and if, for the associated applied stress vector, we write

$$T^\alpha(\eta,t) = Z^\alpha(\eta,z), \quad T(\eta,t) = Z^3(\eta,z) \tag{3.9a,b}$$

we have the notation necessary for the statement of the transformed boundary value problem.

Corresponding to the equilibrium conditions (2.7) and (2.8), we have the pair of differential equations

$$h\sigma^{\beta\alpha}|_\beta + \tau^\alpha,_t = 0 \tag{3.10a}$$

$$h\tau^\alpha|_\alpha + \sigma,_t \quad = 0 \tag{3.10b}$$

together with the single symmetry condition

$$\sigma^{\alpha\beta} = \sigma^{\beta\alpha} \tag{3.11}$$

We have achieved a more consistent appearance by a permutation of the indices in the first equation.

If we combine the strain-displacement relations (2.12) with the constitutive equations (2.14) then, in the transformed notation, we obtain the system of stress-displacement relations

$$w,_t = \frac{h}{E_z}\sigma - \nu_* \frac{h}{E}\sigma^\lambda_\lambda \tag{3.12a}$$

$$u_{\alpha,t} + hw,_\alpha = 2\frac{h}{B}\tau_\alpha \tag{3.12b}$$

$$\frac{1}{2}(u_\alpha|_\beta + u_\beta|_\alpha) = \frac{(1+\nu)}{E}\sigma_{\alpha\beta} - g_{\alpha\beta}\frac{1}{E}[\nu\sigma^\lambda_\lambda + \nu_*\sigma] \tag{3.12c}$$

Recalling relations (2.19) and (2.10*), the formulae for the rotation components

$$\psi = \frac{1}{2}(u_{1,2} - u_{2,1}) \ , \ \omega_\alpha = \frac{1}{B}\tau_\alpha - w,_\alpha \qquad \textbf{(3.13a,b)}$$

follow directly from the transformation (3.7).

In place of the boundary conditions (2.21) and (2.22), we have for the lower and upper faces, respectively,

$$t = -1, \ Y(\overset{\lambda}{x}) \le 0: \ \overset{\alpha}{\tau}(\overset{\alpha}{x}, -1) = 0, \ \sigma(\overset{\lambda}{x}, -1) = p^-(\overset{\lambda}{x}) \qquad \textbf{(3.14a,b)}$$

and

$$t = +1, \ Y(\overset{\lambda}{x}) \le 0: \ \overset{\alpha}{\tau}(\overset{\alpha}{x}, +1) = 0, \ \sigma(\overset{\lambda}{x}, +1) = p^+(\overset{\lambda}{x}) \qquad \textbf{(3.15a,b)}$$

which, together with the transformed edge conditions,

$$-1 \le t \le \pm 1, \ \overset{\lambda}{x} = \overset{\lambda}{x}(\eta): \begin{cases} \overline{\overset{\beta\alpha}{\sigma}(\eta, t)}\ \mathfrak{n}_\beta(\eta) = \overset{\alpha}{T}(\eta, t) & \textbf{(3.16a)} \\ \overline{\overset{\alpha}{\tau}(\eta, t)}\ \mathfrak{n}_\alpha(\eta) = T(\eta, t) & \textbf{(3.16b)} \end{cases}$$

corresponding to (2.23), complete the statement in normalized form.

The boundary value problem now consists of the determination of the solution to the system of equations (3.10) to (3.12) valid in the region $\mathfrak{R}_0$ of (3.3), and satisfying the conditions (3.14) to (3.16).

4. Stress Representations: The Face Conditions

The first step towards recasting the problem in a two dimensional form involves the integration of the equilibrium equations with respect to the normalized transverse variable. Concurrent with the performance of this integration, we derive representations for the stress quantities in which the transverse variation is explicitly separated from the implicit dependence on the laminar coordinates. The imposition of boundary conditions on the faces then leads to the equations to be satisfied by the lower coefficients in these representations.

We introduce an expansion for the second order laminar stress tensor $\overset{\alpha\beta}{\sigma}$, in terms of Legendre polynomial functions of the thickness coordinate, into the equilibrium equations (3.10a) and (3.11). The latter gives the symmetry conditions on the coefficients while in the former the application of a recursion formula for Legendre polynomials effects the t-integration yielding the Legendre representations for the first order tensor of transverse shear stresses. By repeating the operation for

(3.10b), we obtain the corresponding Legendre series for the transverse normal stress.

For these derived series the procedure yields expressions for all coefficients, except the zero-th, in terms of derivatives of the coefficients in the primary expansion for the laminar stress tensor. The unspecified zero-th coefficients are the unknown functions introduced by the successive integrations: those appearing at the first step constitute a first order tensor while the second integration adds a single unknown function of the laminar coordinates. The fulfillment of the requirements on the faces leads to the set of relations linking these quantities to the leading coefficients in the primary expansions.

We shall see that the relations representing the satisfaction of the face conditions, when combined in an appropriate manner, yield respectively

(1) The system of partial differential equations to be satisfied by the components of the zero-th laminar stress coefficient tensor.

(2) The partial differential equation to be satisfied by the components of the first laminar stress coefficient tensor.

(3) A tensor relation expressing the zero-th transverse shear stress coefficient tensor in terms of the derivative of the first laminar stress coefficient tensor.

(4) An expression for the zero-th transverse normal stress coefficient in terms of derivatives of the second laminar stress coefficient tensor.

In fact the first three can be interpreted as the equlibrium conditions of the classical two-dimensional theories. Observing that the zero-th and first laminar stress coefficient tensors correspond respectively to the stress resultant and stress couple tensors, and further noting the relations between the tensor of transverse shear resultants and the zero-th transverse shear coefficient tensor, it will be evident that the first system consists of the classical equilibrium conditions for generalized plane stress while the system composed of the second and third combined is equivalent to the equilibrium conditions of plate bending.

For the integration of the Legendre series we shall make repeated use of a recursion formula for Legendre polynomials. Letting $P_n(t)$ denote the Legendre polynomial in t of degree n and using a dot to denote d/dt we recall that*

$$(2n+1)\,P_n(t) = \dot{P}_{n+1}(t) - \dot{P}_{n-1}(t) \qquad \textbf{(4.1)}$$

The integrated form of this relation which, apart from an additive term independent of t, reads

$$\int P_n(t)\,dt = \frac{1}{2n+1}\,[P_{n+1}(t) - P_{n-1}(t)] \qquad \textbf{(4.2)}$$

will be used repeatedly in our development

We write the expansion for the laminar stress tensor in the form

$$\sigma^{\alpha\beta}(x^\lambda,t) = \sum_{n=0}^{\infty} \overset{(n)}{s}{}^{\alpha\beta}(x^\lambda)\,P_n(t) \qquad \textbf{(4.3)}$$

* See e.g. §15.21 of Modern Analysis by E. T. Whittaker and G. N. Watson, C.U.P., Cambridge, 1952.

in which the coefficients $\overset{(n)}{S}{}^{\alpha\beta}$ are undetermined functions of the two laminar coordinates x^λ. The symmetry condition (3.10) implies the symmetry of each coefficient, namely

$$\overset{(n)}{S}{}^{\alpha\beta} = \overset{(n)}{S}{}^{\beta\alpha}, \quad \alpha,\beta = 1,2, \; n = 0,1,2\ldots\infty. \tag{4.4}$$

If we introduce the expansion (4.3) into the first equilibrium equation (3.10a) and make a transposition, we have

$$\tau^\alpha{}_{,t} = -h \sum_{n=0}^{\infty} \overset{(n)}{S}{}^{\beta\alpha}|_\beta \cdot P_n \tag{4.5}$$

in which we no longer exhibit explicitly the respective dependence on the coordinates. The integration of equation (4.5) follows from the application of formula (4.2): after a rearrangement of terms, we obtain the expansion

$$\tau^\alpha = \sum_{n=0}^{\infty} \overset{(n)}{\tau}{}^\alpha (x^\lambda)\, P_n(t) \tag{4.6}$$

in which $\overset{(0)}{\tau}{}^\alpha$ is an undetermined first order tensor introduced by the integration, and the higher coefficients are given by

$$\overset{(n)}{\tau}{}^\alpha = h \left[\frac{\overset{(n+1)}{S}{}^{\beta\alpha}|_\beta}{2n+3} - \frac{\overset{(n-1)}{S}{}^{\beta\alpha}|_\beta}{2n-1} \right], \; n \geq 1 \tag{4.7}$$

Proceeding to the second equilibrium equation, we introduce the expansion (4.6) into (3.10b) and, transposing, we obtain

$$\sigma_{,t} = -h \sum_{n=0}^{\infty} \overset{(n)}{\tau}{}^\alpha|_\alpha P_n \tag{4.8}$$

Performing the integration in the same manner as before with the aid of formula (4.2), we find for the transverse normal stress

$$\sigma = \sum_{n=0}^{\infty} \overset{(n)}{\sigma} (x^\lambda)\, P_n(t) \tag{4.9}$$

in which $\overset{(0)}{\sigma}$ is an unknown function introduced by the integration and, the higher coefficients are given by

$$\overset{(n)}{\sigma} = h \left[\frac{\overset{(n+1)}{\tau}{}^\alpha|_\alpha}{2n+3} - \frac{\overset{(n-1)}{\tau}{}^\alpha|_\alpha}{2n-1} \right], \; n \geq 1 \tag{4.10}$$

By substituting for the $\overset{(n)}{\tau}{}^{\alpha}$ from (4.7) into (4.10), we see that, for $n \geq 2$, the $\overset{(n)}{\sigma}$ are related to the laminar coefficients $\overset{(n)}{S}{}^{\alpha\beta}$ by

$$\overset{(n)}{\sigma} = h^2 \left[\frac{\overset{(n-2)}{S}{}^{\alpha\beta}|_{\beta}}{(2n-3)(2n-1)} - \frac{2\,\overset{(n)}{S}{}^{\alpha\beta}|_{\alpha\beta}}{(2n-1)(2n+3)} + \frac{\overset{(n+2)}{S}{}^{\alpha\beta}|_{\alpha\beta}}{(2n+3)(2n+5)} \right], \quad n \geq 2 \qquad \textbf{(4.11)}$$

while the excluded formula for $\overset{(1)}{\sigma}$ may be written

$$\overset{(1)}{\sigma} = h^2 \left[\frac{1}{35}\overset{(3)}{S}{}^{\alpha\beta}|_{\alpha\beta} - \frac{1}{15}\overset{(1)}{S}{}^{\alpha\beta}|_{\alpha\beta} \right] - h\,\overset{(0)}{\tau}{}^{\alpha}|_{\alpha} \qquad \textbf{(4.12)}$$

in which $\overset{(0)}{\tau}{}^{\alpha}$ is still undetermined.

Turning now to the application of the face conditions, we recall that, for Legendre polynomials, we have the terminal values

$$P_n(-1) = (-1)^n, \quad P_n(1) = 1 \qquad \textbf{(4.13)}$$

which, when used in the expressions (4.6) and (4.9), give respectively

$$\tau^{\alpha}(x^{\lambda}, -1) = \overset{(0)}{\tau}{}^{\alpha} + \sum_{n=1}^{\infty} (-1)^n \overset{(n)}{\tau}{}^{\alpha}, \quad \tau^{\alpha}(x^{\lambda}, +1) = \overset{(0)}{\tau}{}^{\alpha} + \sum_{n=1}^{\infty} \overset{(n)}{\tau}{}^{\alpha} \; . \qquad \textbf{(4.14a,b)}$$

and

$$\sigma(x^{\lambda}, -1) = \overset{(0)}{\sigma} + \sum_{n=1}^{\infty} (-1)^n \overset{(n)}{\sigma}, \quad \sigma(x^{\lambda}, +1) = \overset{(0)}{\sigma} + \sum_{n=1}^{\infty} \overset{(n)}{\sigma} \qquad \textbf{(4.15a,b)}$$

If we insert formulae (4.7) for the $\overset{(n)}{\tau}{}^{\alpha}$ into (4.14), and note the cancellations resulting from the recursive feature of (4.7), we find

$$\tau^{\alpha}(x^{\lambda}, -1) = \overset{(0)}{\tau}{}^{\alpha} + h\,\overset{(0)}{S}{}^{\beta\alpha}|_{\beta} - \frac{1}{3} h\,\overset{(1)}{S}{}^{\beta\alpha}|_{\beta} \qquad \textbf{(4.16a)}$$

$$\tau^{\alpha}(x^{\lambda}, +1) = \overset{(0)}{\tau}{}^{\alpha} - h\,\overset{(0)}{S}{}^{\beta\alpha}|_{\beta} - \frac{1}{3} h\,\overset{(1)}{S}{}^{\beta\alpha}|_{\beta} \qquad \textbf{(4.16b)}$$

Similarly, by introducing (4.10) into (4.15) and making the corresponding cancellations, we obtain

$$\sigma(x^\lambda, -1) = \overset{(0)}{\sigma} + h\, \overset{(0)}{\tau}{}^{\alpha}|_\alpha - \frac{1}{3} h\, \overset{(1)}{\tau}{}^{\alpha}|_\alpha \tag{4.17a}$$

$$\sigma(x^\lambda, +1) = \overset{(0)}{\sigma} - h\, \overset{(0)}{\tau}{}^{\alpha}|_\alpha - \frac{1}{3} h\, \overset{(1)}{\tau}{}^{\alpha}|_\alpha \tag{4.17b}$$

The above relations enable us to write the boundary conditions on the faces in terms of the coefficients appearing on the right of (4.16) and (4.17). The face conditions (3.14a) and (3.15a) then become

$$\overset{(0)}{\tau}{}^{\alpha} + h\, \overset{(0)}{S}{}^{\beta\alpha}|_\beta - \frac{1}{3} h\, \overset{(1)}{S}{}^{\beta\alpha}|_\beta = 0 \tag{4.18a}$$

and

$$\overset{(0)}{\tau}{}^{\alpha} - h\, \overset{(0)}{S}{}^{\beta\alpha}|_\beta - \frac{1}{3} h\, \overset{(1)}{S}{}^{\beta\alpha}|_\beta = 0 \tag{4.18b}$$

respectively; correspondingly, the conditions (3.14b) and (3.15b) take the respective forms

$$\overset{(0)}{\sigma} + h\, \overset{(0)}{\tau}{}^{\alpha}|_\alpha - \frac{1}{3} h\, \overset{(1)}{\tau}{}^{\alpha}|_\alpha = p^- \tag{4.19a}$$

and

$$\overset{(0)}{\sigma} - h\, \overset{(0)}{\tau}{}^{\alpha}|_\alpha - \frac{1}{3} h\, \overset{(1)}{\tau}{}^{\alpha}|_\alpha = p^+ \tag{4.19b}$$

The addition and subtraction of relations (4.18) yield, respectively,

$$\overset{(0)}{S}{}^{\beta\alpha}|_\beta = 0 \tag{4.20}$$

and

$$\overset{(0)}{\tau}{}^{\alpha} = \frac{1}{3} h\, \overset{(1)}{S}{}^{\beta\alpha}|_\beta \tag{4.21}$$

while similar combinations of relations (4.19) give

$$2h\, \overset{(0)}{\tau}{}^{\alpha}|_\alpha + p = 0 \tag{4.22}$$

and

$$\overset{(0)}{\sigma} = \frac{1}{3} h \overset{(1)}{\tau}{}^{\alpha} |_{\alpha} + \frac{1}{2} p^{*} \tag{4.23}$$

respectively, where we have used the notation

$$p^{*} = \overset{+}{p} + \overset{-}{p}, \quad p = \overset{+}{p} - \overset{-}{p} \tag{4.24}$$

for the pinching and bending effects of the surface pressures.

Relation (4.21) expressing the tensor $\overset{(0)}{\tau}{}^{\alpha}$ in terms of the first laminar coefficient tensor furnishes the formula for the completion of the set of relations (4.7): similarly, formula (4.23) for $\overset{(0)}{\sigma}$ provides the missing member of the set (4.10).

As the components of the tensor $\overset{(0)}{\tau}{}^{\alpha}$ must satisfy equation (4.22), we may substitute from (4.21) and obtain the equivalent condition in terms of the first laminar stress coefficient tensor, namely

$$\frac{2}{3} h^{2} \overset{(1)}{S}{}^{\alpha\beta} |_{\alpha\beta} + p = 0 \tag{4.25}$$

Thus, in relations (4.20) and (4.25), we have the equations to be satisfied by the zero-th and first coefficients in the expansion for the laminar stress tensor.

Besides the expression (4.21) for $\overset{(0)}{\tau}{}^{\alpha}$, we also list explicitly the formulae for the next two coefficients in the expansion for the transverse shear stress. Noting equation (4.20), we see that (4.7) with $n = 1$ gives

$$\overset{(1)}{\tau}{}^{\alpha} = \frac{1}{5} h \overset{(2)}{S}{}^{\beta\alpha} |_{\beta} \tag{4.26}$$

while the case $n = 2$ has the standard form

$$\overset{(2)}{\tau}{}^{\alpha} = h \left[\frac{1}{7} \overset{(3)}{S}{}^{\beta\alpha} |_{\beta} - \frac{1}{3} \overset{(1)}{S}{}^{\beta\alpha} |_{\beta} \right] \tag{4.27}$$

and it is evident from (4.7) that the formulae for the $\overset{(n)}{\tau}{}^{\alpha}$ with n greater than two involve only those laminar stress coefficients with index n at least two.

For the relations excluded from (4.11), we first introduce relation (4.26) into formula (4.23) and obtain

$$\overset{(0)}{\sigma} = \frac{1}{15} h^{2} \overset{(2)}{S}{}^{\alpha\beta} |_{\alpha\beta} + \frac{1}{2} p^{*} \tag{4.28}$$

while the use of relations (4.22) and (4.25) in formula (4.12) yields

$$\overset{(1)}{\sigma} = \frac{1}{35} h^2 \overset{(3)}{S}{}^{\alpha\beta}\,|_{\alpha\beta} + \frac{3}{5} p \tag{4.29}$$

For later convenience, we also list explicitly the next two coefficients as given by (4.11): in the case $n = 2$, if we note relation (4.20), we have

$$\overset{(2)}{\sigma} = h^2 \left[-\frac{2}{21} \overset{(2)}{S}{}^{\alpha\beta}\,|_{\alpha\beta} + \frac{1}{63} \overset{(4)}{S}{}^{\alpha\beta}\,|_{\alpha\beta} \right] \tag{4.30}$$

while using equation (4.25) in the formula for $\overset{(3)}{\sigma}$ gives

$$\overset{(3)}{\sigma} = h^2 \left[-\frac{2}{45} \overset{(3)}{S}{}^{\alpha\beta}\,|_{\alpha\beta} + \frac{1}{99} \overset{(5)}{S}{}^{\alpha\beta}\,|_{\alpha\beta} \right] - \frac{1}{10} p \tag{4.31}$$

and we note that the formulae for the higher coefficients ($n \geq 4$) involve only those laminar coefficients with index n at least two.

In deriving Legendre series for the transverse shear and normal stresses, in which the coefficients are expressed as derivatives of the coefficients appearing in the original expansion for the laminar stresses, we have also extracted the equations to be satisfied by the zero-th and first laminar stress coefficients: the equations for the determination of the higher coefficients in the original expansion will be deduced at a later stage. Here we merely observe that in relations (4.28) to (4.31) and in the general formula (4.11), we have expressions for the transverse normal stress coefficients in which there occurs only those coefficients with index at least two in the original laminar stress expansion, together with the applied surface forces.

Equations (4.20) and (4.21) for the zero-th and first laminar stress coefficients and equation (4.22) for the zero-th transverse shear coefficient are the integrated equilibrium equations of the classical theories. If we define the stress resultants $N^{\alpha\beta}$ and Q^{α} and stress couple $M^{\alpha\beta}$ by

$$N^{\alpha\beta} = \int_{-h}^{h} \tau^{\alpha\beta}\, dz = h \int_{-1}^{1} \sigma^{\alpha\beta}\, dt = 2h \overset{(0)}{S}{}^{\alpha\beta} \tag{4.32a}$$

$$M^{\alpha\beta} = \int_{-h}^{h} z\, \tau^{\alpha\beta}\, dz = h^2 \int_{-1}^{1} t\, \sigma^{\alpha\beta}\, dt = \frac{2}{3} h^2 \overset{(1)}{S}{}^{\alpha\beta} \tag{4.32b}$$

$$Q^{\alpha} = \int_{-h}^{h} \tau^{\alpha 3}\, dz = h \int_{-1}^{1} \tau^{\alpha}\, dt = 2h \overset{(0)}{\tau}{}^{\alpha} \tag{4.32c}$$

then, in this alternative notation, equation (4.20) takes the more familiar form

$$N^{\beta\alpha}\,|_{\beta} = 0 \tag{4.33}$$

describing equilibrium in the generalized plane stress version of the stretching problem, while equations (4.21) and (4.22) become the classical equilibrium condition for plate bending, namely

$$Q^{\alpha} = M^{\beta\alpha}\,|_{\beta}, \quad Q^{\alpha}\,|_{\alpha} + p = 0 \tag{4.34a,b}$$

The elimination of Q^{α} from the latter system yields the alternate form

$$M^{\alpha\beta}\,|_{\alpha\beta} + p = 0 \tag{4.35}$$

equivalent to equation (4.25).

5. The Edge Boundary Conditions

Having derived the Legendre representations for the stress components, we now effect an appropriate reformulation of the conditions on the edge surface. The edge conditions (3.16a) on the laminar stresses are equivalent to a sequence of conditions on the laminar coefficients $\overset{(n)}{S}{}^{\alpha\beta}$: Similarly, the conditions (3.16b) on the shear stresses are transformed into a corresponding set of conditions on the shear coefficients $\overset{(n)}{\tau}{}^{\alpha\beta}$.

We have already noted that the application of the face conditions to the stress representation led to the explicit equations (4.20), (4.21) and (4.22) for the initial coefficients. The conditions we derive on the coefficients $\overset{(0)}{S}{}^{\alpha\beta}$ are associated with the system of equations (4.20) describing the stretching problem, while the conditions we deduce for the coefficients $\overset{(1)}{S}{}^{\alpha\beta}$ and $\overset{(0)}{\tau}{}^{\alpha}$ complement equations (4.21) and (4.22) for the problem of bending. By relating the edge conditions to their respective equations, we can infer the constraints on the initial coefficients of the applied edge stress distributions necessary to ensure overall static equilibrium.

The conditions on the remaining coefficients $\{\overset{(n)}{S}{}^{\alpha\beta},\ n \geq 2:\ \overset{(n)}{\tau}{}^{\alpha},\ n \geq 1\}$ will be associated with the system of equations for the higher stress coefficients to be derived later. As these latter equations will be formulated entirely in terms of the laminar coefficients, it is desirable to have all edge conditions described in the same terms. Accordingly, we transform the set of conditions on the $\overset{(n)}{\tau}{}^{\alpha}$ into an equivalent set expressing a further series of conditions on the $\overset{(n)}{S}{}^{\alpha\beta}$.

Writing for the edge values of the coefficients

$$\overline{\overset{(n)}{S}{}^{\alpha\beta}}(\eta) = \left[\overset{(n)}{S}{}^{\alpha\beta}(x^{\lambda})\right]_{x^{\lambda}=\overline{x^{\lambda}(\eta)}}, \quad \overline{\overset{(n)}{\tau}{}^{\alpha}}(\eta) = \left[\overset{(n)}{\tau}{}^{\alpha}(x^{\lambda})\right]_{x^{\lambda}=\overline{x^{\lambda}(\eta)}} \tag{5.1a,b}$$

the edge values (3.8) of the stress components have the representations

$$\overline{\sigma^{\alpha\beta}}(\eta,t) = \sum_{n=0}^{\infty} \overline{\overset{(n)}{S}{}^{\alpha\beta}}(\eta)\, P_n(t), \quad \overline{\tau^{\alpha}}(\eta,t) = \sum_{n=0}^{\infty} \overline{\overset{(n)}{\tau}{}^{\alpha}}(\eta)\, P_n(t) \tag{5.2a,b}$$

Hence, if we expand the components of the applied edge stress distribution in the form

$$T^{\alpha}(\eta,t) = \sum_{n=0}^{\infty} \overset{(n)}{T}{}^{\alpha}(\eta)\, P_n(t), \quad T(\eta,t) = \sum_{n=0}^{\infty} \overset{(n)}{T}(\eta)\, P_n(t) \tag{5.3a,b}$$

it follows from the orthogonality of the Legendre polynomials that the edge conditions (3.16) are equivalent to the double sequence of conditions on the coefficients

$$x^{\lambda} = \overline{x^{\lambda}(\eta)}: \begin{cases} \overline{\overset{(n)}{S}{}^{\alpha\beta}}(\eta)\, \mathrm{n}_{\beta}(\eta) = \overset{(n)}{T}{}^{\alpha}(\eta), \; n \geq 0 & \text{(5.4a)} \\ \overline{\overset{(n)}{\tau}{}^{\alpha}}(\eta)\, \mathrm{n}_{\alpha}(\eta) = \overset{(n)}{T}(\eta), \; n \geq 0 & \text{(5.4b)} \end{cases}$$

In particular we have the conditions

$$\overline{\overset{(0)}{S}{}^{\beta\alpha}}(\eta)\, \mathrm{n}_{\beta}(\eta) = \overset{(0)}{T}{}^{\alpha}(\eta) \tag{5.5}$$

associated with set of equations (4.20), while the edge requirements

$$\overline{\overset{(1)}{S}{}^{\beta\alpha}}(\eta)\, \mathrm{n}_{\beta}(\eta) = \overset{(1)}{T}{}^{\alpha}(\eta), \quad \overline{\overset{(0)}{\tau}{}^{\alpha}}(\eta)\, \mathrm{n}_{\alpha}(\eta) = \overset{(0)}{T}(\eta) \tag{5.6a,b}$$

complement the system of equations (4.21) and (4.22) combined.

In order to transform conditions (5.4b) on the tensor coefficients $\overset{(n)}{\tau}{}^{\alpha}$ into an equivalent sequence of conditions on the coefficients $\overset{(n)}{S}{}^{\alpha\beta}$, we use relations (4.7) and note the particular formulae (4.21) and (4.26). Starting with expression (4.21) for $\overset{(0)}{\tau}{}^{\alpha}$, we then use (4.7) to write successively the formulae for the $\overset{(n)}{\tau}{}^{\alpha}$ with even index n: thus, we obtain the sequence of relations for the $\overset{(2k)}{\tau}{}^{\alpha}$

$$\overset{(0)}{\tau}{}^{\alpha} = \frac{1}{3} h \overset{(1)}{S}{}^{\beta\alpha}|_{\beta}$$

$$\overset{(2)}{\tau}{}^{\alpha} = \frac{1}{7} h \overset{(3)}{S}{}^{\beta\alpha}|_{\beta} - \frac{1}{3} h \overset{(1)}{S}{}^{\beta\alpha}|_{\beta}$$

$$\overset{(4)}{\tau}{}^{\alpha} = \frac{1}{11} h \overset{(5)}{S}{}^{\beta\alpha}|_{\beta} - \frac{1}{7} h \overset{(3)}{S}{}^{\beta\alpha}|_{\beta}$$

$$\vdots \qquad\qquad \ddots \qquad\qquad \ddots$$

$$\overset{(2k)}{\tau}{}^{\alpha} = \frac{1}{4k+3} h \overset{(2k+1)}{S}{}^{\beta\alpha}|_{\beta} - \frac{1}{4k-1} h \overset{(2k-1)}{S}{}^{\beta\alpha}|_{\beta}$$

which, when added, give for $k \geq 0$,

$$\sum_{i=0}^{k} \overset{(2i)}{\tau}{}^{\alpha} = \frac{1}{4k+3} h \overset{(2k+1)}{S}{}^{\beta\alpha}|_{\beta}, \; k \geq 0 \tag{5.7}$$

or alternatively,

$$h \overset{(2k+1)}{S}{}^{\beta\alpha}|_{\beta} = (4k+3) \sum_{i=0}^{k} \overset{(2i)}{\tau}{}^{\alpha}, \; k \geq 0 \tag{5.8}$$

Similarly, we may start with expression (4.26) for $\overset{(1)}{\tau}{}^{\alpha}$ and then use (4.7) to list successively the formulae for the $\overset{(n)}{\tau}{}^{\alpha}$ with odd index n: performing a corresponding summation on these relations for the $\overset{(2k+1)}{\tau}{}^{\alpha}$, we find for $k \geq 1$,

$$\sum_{i=0}^{(k-1)} \overset{(2i+1)}{\tau}{}^{\alpha} = \frac{1}{4k+1} h \overset{(2k)}{S}{}^{\beta\alpha}|_{\beta}, \; k \geq 1 \tag{5.9}$$

or alternatively,

$$h \overset{(2k)}{S}{}^{\beta\alpha}|_{\beta} = (4k+1) \sum_{i=0}^{(k-1)} \overset{(2i+1)}{\tau}{}^{\alpha}, \; k \geq 1 \tag{5.10}$$

The system of relations (5.7) is the generalization for the higher coefficients of formula (4.21), while relations (4.20) correspond to the case $k = 0$ of (5.9).

We shall use relations (5.8) and (5.10) only at their edge values for which we have

$$\overline{h\overset{(2k+1)}{S}{}^{\beta\alpha}|_\beta} \equiv \left[h\overset{(2k+1)}{S}{}^{\beta\alpha}|_\beta\right]_{\overline{x^\lambda = x^\lambda(\eta)}} = (4k+3)\sum_{i=0}^{k}\overline{\overset{(2i)}{\tau}{}^{\alpha}(\eta)} \tag{5.11a}$$

$$\overline{h\overset{(2k)}{S}{}^{\beta\alpha}|_\beta} \equiv \left[h\overset{(2k)}{S}{}^{\beta\alpha}|_\beta\right]_{\overline{x^\lambda = x^\lambda(\eta)}} = (4k+1)\sum_{i=0}^{(k-1)}\overline{\overset{(2i+1)}{\tau}{}^{\alpha}(\eta)} \tag{5.11b}$$

If we set

$$\overset{(2k+1)}{D}(\eta) = (4k+3)\sum_{i=0}^{k}\overset{(2i)}{T}(\eta),\ k \geq 0 \tag{5.12a}$$

$$\overset{(2k)}{D}(\eta) = (4k+1)\sum_{i=0}^{(k-1)}\overset{(2i+1)}{T}(\eta),\ k \geq 1 \tag{5.12b}$$

we see that the system of edge conditions (5.4) is equivalent to

$$x^\lambda = \overline{x^\lambda(\eta)}: \begin{cases} \overline{\overset{(n)}{S}{}^{\beta\alpha}}\, n_\beta = \overset{(n)}{T}{}^{\alpha}(\eta),\ n \geq 0 & (5.13a) \\ \overline{h\overset{(n)}{S}{}^{\beta\alpha}|_\beta}\, n_\alpha = \overset{(n)}{D}(\eta),\ n \geq 1 & (5.13b) \end{cases}$$

which, for convenience in later reference, we separate into the three sets

$$\overline{\overset{(0)}{S}{}^{\beta\alpha}}\, n_\beta = \overset{(0)}{T}{}^{\alpha}(\eta) \tag{5.14a}$$

$$\overline{\overset{(1)}{S}{}^{\beta\alpha}}\, n_\beta = \overset{(1)}{T}{}^{\alpha},\quad \overline{h\overset{(1)}{S}{}^{\beta\alpha}|_\beta}\, n_\alpha = \overset{(1)}{D}(\eta) \tag{5.14b}$$

$$\overline{\overset{(n)}{S}{}^{\beta\alpha}}\, n_\beta = \overset{(n)}{T}{}^{\alpha},\quad \overline{h\overset{(n)}{S}{}^{\beta\alpha}|_\beta}\, n_\alpha = \overset{(n)}{D}(\eta),\ n \geq 2 \tag{5.14c}$$

The set (5.14a), identical with (5.5) is associated with the system of equations (4.20). The second group (5.14b), equivalent to conditions (5.6), gives the set of conditions to be assigned to the differential equation (4.25). The sequence (5.14c) constitute the set of edge conditions complementing the system of differential equations for the highest laminar coefficients $\{\overset{(n)}{S}{}^{\alpha\beta}, n \geq 2\}$ to be formulated later.

By considering the integral over the edge surface of the applied edge laminar stresses, and integrating through the thickness, we find

$$\iint_{\mathcal{B}} Z^{\alpha}\, d\,\mathcal{B} = 2h \oint_{\mathcal{C}} \overset{(0)}{T}{}^{\alpha}\, d\,\mathcal{C} = 2h \oint_{\mathcal{C}} \overset{(0)}{S}{}^{\beta\alpha}\, \mathfrak{n}_{\beta}\, d\,\mathcal{C} \tag{5.15}$$

in which the latter integrals are taken over the bounding curve and where we have used (5.14a). If we now apply Gauss' Theorem, we can express the line integral on the right as an integral over the area occupied by the planar projection of the plate; we then have

$$2h \oint_{\mathcal{C}} \overset{(0)}{T}{}^{\alpha}\, d\,\mathcal{C} = 2h \iint_{\mathcal{A}} \overset{(0)}{S}{}^{\beta\alpha}\,|_{\beta}\, d\,\mathcal{A} \tag{5.16}$$

which vanishes by virtue of (4.20), so that we have the constraint

$$\oint_{\mathcal{C}} \overset{(0)}{T}{}^{\alpha}\, d\,\mathcal{C} = 0 \tag{5.17}$$

on the zero-th coefficient of the applied edge stress. Similarly, considering the second of relations (5.14b), and again using Gauss' Theorem, we have

$$\frac{2}{3}h \oint_{\mathcal{C}} \overset{(1)}{D}\,(\eta)\, d\mathcal{C} = \frac{2}{3}h^{2} \oint_{\mathcal{C}} \overset{(1)}{S}{}^{\beta\alpha}\,|_{\beta}\, \mathfrak{n}_{\beta}\, d\mathcal{C} = \frac{2}{3}h^{2} \iint_{\mathcal{A}} \overset{(1)}{S}{}^{\beta\alpha}\,|_{\beta\alpha}\, d\mathcal{A} \tag{5.18}$$

which, by virtue of (4.25), gives

$$\frac{2}{3}h \oint_{\mathcal{C}} \overset{(1)}{D}\, d\,\mathcal{C} + \iint_{\mathcal{A}} p\, d\mathcal{A} = 0 \tag{5.19}$$

as a further constraint necessary for overall static equilibrium.

6. Representations for the Displacement Components

As the integration of the equilibrium equations (3.10) led to the Legendre series for the stresses, so shall the integration of the constitutive relations (3.12a,b) yield the corresponding representations for the displacement components. Recalling our restriction to elastically homogeneous media, implying constancy for the elastic parameters, we introduce the fixed dimensionless ratios Λ_* and Λ, respectively associated with the extensibility and shear deformability in the transverse direction, and defined by

$$\Lambda_* = \frac{E}{E_z}, \quad \Lambda(1+\nu) = \frac{E}{B} \tag{6.1a,b}$$

so that relations (3.12a,b) may now be written

$$w,_t = \frac{h}{E}\left[\Lambda_* \sigma - \nu_* \sigma^\lambda_\lambda\right] \tag{6.2a}$$

$$u_{\alpha,t} = 2\,\frac{h}{E}\,\Lambda(1+\nu)\,\tau_\alpha - h w,_\alpha \tag{6.2b}$$

where, in the latter, we have transposed the transverse displacement term. In terms of the new parameters isotropy is characterized by

$$\Lambda_* = \Lambda = 1, \; \nu_* = \nu \tag{6.3}$$

while for transverse inextensibility, we would have

$$\Lambda_* = \nu_* = 0 \tag{6.4}$$

Formula (3.13b) for the transverse components of infinitesimal rotation now reads

$$\omega_\alpha = \frac{1}{E}\,\Lambda(1+\nu)\,\tau_\alpha - w,_\alpha \tag{6.5a}$$

while formula (3.13a) is unaffected so that we still have

$$\psi = \frac{1}{2}\left(u_{1,2} - u_{2,1}\right) \tag{6.5b}$$

for the laminar rotation.

If the expansion (4.3) for $\sigma^{\alpha\beta}$ together with the related representation (4.9) for σ are introduced into (6.2a), there follows the series form of the latter equation, namely

$$w,_t = \frac{h}{E}\sum_{n=0}^{\infty}\left(\Lambda_* \overset{(n)}{\sigma} - \nu_* \overset{(n)}{S}{}^\lambda_\lambda\right) P_n \tag{6.6}$$

The application of the integration formula (4.2) immediately yields the Legendre series for the transverse displacement in the form

$$w = \sum_{n=0}^{\infty} \overset{(n)}{w}(x^\lambda)\, P_n(t) \tag{6.7}$$

where $\overset{(0)}{w}$ is an unknown function introduced by the integration and the higher coefficients are given by

$$\overset{(n)}{w} = -\frac{h}{E}\left[\frac{\Lambda_* \overset{(n+1)}{\sigma} - \nu_* \overset{(n+1)}{S}{}^\lambda_\lambda}{2n+3} - \frac{\Lambda_* \overset{(n-1)}{\sigma} - \nu_* \overset{(n-1)}{S}{}^\lambda_\lambda}{2n-1}\right], \quad n \ge 1 \tag{6.8}$$

We next write the expansions for the laminar derivative of the transverse displacement which appears both in equation (6.2b) and in formula (6.5a). Since, for invariant quantities, covariant differentiation is indistinguishable from partial differentiation, we shall use a stroke rather than a comma in writing the partial derivatives of the terms in the coefficients (6.8): this avoids the necessity of making an adjustment when we come to taking second derivatives later: we obtain

$$w_{,\alpha} = \overset{(0)}{w}|_\alpha - \frac{h}{E}\sum_{n=1}^{\infty}\left[\frac{\Lambda_* \overset{(n+1)}{\sigma}|_\alpha - \nu_* \overset{(n+1)}{S}{}^{\lambda}{}_{\lambda|\alpha}}{2n+3} - \frac{\Lambda_* \overset{(n-1)}{\sigma}|_\alpha - \nu_* \overset{(n-1)}{S}{}^{\lambda}{}_{\lambda|\alpha}}{2n-1}\right] P_n \tag{6.9}$$

Moreover, recalling relations (4.21), (4.26), (4.27) and the general form (4.7) in connection with the series (4.6) for the transverse shear stress, and relabelling the dummy index, we have the explicit expansion for τ_α

$$\tau_\alpha = \frac{1}{3}\, h \overset{(1)}{S}{}_{\alpha}{}^{\lambda}{}_{|\lambda} + \frac{1}{5}\, h \overset{(2)}{S}{}_{\alpha}{}^{\lambda}{}_{|\lambda} P_1 + h \sum_{n=2}^{\infty}\left[\frac{\overset{(n+1)}{S}{}_{\alpha}{}^{\lambda}{}_{|\lambda}}{2n+3} - \frac{\overset{(n-1)}{S}{}_{\alpha}{}^{\lambda}{}_{|\lambda}}{2n-1}\right] P_n \tag{6.10}$$

If we combine the series (6.9) and (6.10) in accordance with formula (6.5a), we obtain the Legendre representations for the transverse rotations in the form

$$\omega_\alpha = \sum_{n=0}^{\infty} \overset{(n)}{\omega}(x^\lambda)\, P_n(t) \tag{6.11}$$

with the notation

$$\overset{(0)}{\omega}_\alpha = \frac{h}{E}\Lambda(1+\nu)\,\frac{1}{3}\, h\, \overset{(1)}{S}{}_{\alpha}{}^{\lambda}{}_{|\lambda} - \overset{(0)}{w}|_\alpha \tag{6.12a}$$

$$\overset{(1)}{\omega}_\alpha = \frac{h}{E}\left[\frac{1}{5}\left(\Lambda(1+\nu)\, \overset{(2)}{S}{}_{\alpha}{}^{\lambda}{}_{|\lambda} - \nu_* \overset{(2)}{S}{}^{\lambda}{}_{\lambda|\alpha} + \Lambda_* \overset{(2)}{\sigma}|_\alpha\right) - \left(-\nu_* \overset{(0)}{S}{}^{\lambda}{}_{\lambda|\alpha} + \Lambda_* \overset{(0)}{\sigma}|_\alpha\right)\right] \tag{6.12b}$$

$$\overset{(n)}{\omega}_\alpha = \frac{h}{E}\left[\frac{\Lambda(1+\nu)\, \overset{(n+1)}{S}{}_{\alpha}{}^{\lambda}{}_{|\lambda} - \nu_* \overset{(n+1)}{S}{}^{\lambda}{}_{\lambda|\alpha} + \nu_* \overset{(n+1)}{\sigma}|_\alpha}{2n+3} - \frac{\Lambda(1+\nu)\, \overset{(n-1)}{S}{}_{\alpha}{}^{\lambda}{}_{|\lambda} - \nu_* \overset{(n-1)}{S}{}^{\lambda}{}_{\lambda|\alpha} + \Lambda_* \overset{(n-1)}{\sigma}|_\alpha}{2n-1}\right], \quad n \geq 2 \tag{6.12c}$$

In a similar manner, we can obtain the series form of equation (6.2b) by inserting the expansions (6.9) and (6.10) into the right hand side: we find

$$
\begin{aligned}
u_{\alpha,t} = {} & \left[\frac{h^2}{E} \Lambda(1+\nu) \frac{2}{3} \overset{(1)}{S}{}^{\lambda}_{\alpha|\lambda} - h \overset{(0)}{w}|_{\alpha} \right] P_0 \\
& + \frac{h}{E} \left[\frac{1}{5} \left(2\Lambda(1+\nu) \overset{(2)}{S}{}^{\lambda}_{\alpha|\lambda} - \nu_* \overset{(2)}{S}{}^{\lambda}_{\lambda|\alpha} + \Lambda_* \overset{(2)}{\sigma}|_{\alpha} \right) - \left(-\nu_* \overset{(0)}{S}{}^{\lambda}_{\lambda|\alpha} + \Lambda_* \overset{(0)}{\sigma}|_{\alpha} \right) \right] P_1 \\
& + \frac{h^2}{E} \sum_{n=2}^{\infty} \left[\frac{2\Lambda(1+\nu) \overset{(n+1)}{S}{}^{\lambda}_{\alpha|\lambda} - \nu_* \overset{(n+1)}{S}{}^{\lambda}_{\lambda|\alpha} + \Lambda_* \overset{(n+1)}{\sigma}|_{\alpha}}{2n+3} \right. \\
& \qquad \left. - \frac{2\Lambda(1+\nu) \overset{(n-1)}{S}{}^{\lambda}_{\alpha|\lambda} - \nu_* \overset{(n-1)}{S}{}^{\lambda}_{\lambda|\alpha} + \Lambda_* \overset{(n-1)}{\sigma}|_{\alpha}}{2n-1} \right] P_n
\end{aligned}
\tag{6.13}
$$

the integration of which can be effected by again applying formula (4.2): after a rearrangement, we obtain the Legendre representations for the laminar displacements in the form

$$
u_{\alpha} = \sum_{n=0}^{\infty} \overset{(n)}{u}_{\alpha}(x^{\lambda})\, P_n(t) \tag{6.14}
$$

in which the $\overset{(0)}{u}_{\alpha}$ are the undetermined components introduced by the integration and the higher coefficients are given as follows

$$
\begin{aligned}
\overset{(0)}{u}_{\alpha} = {} & -h \overset{(0)}{w}|_{\alpha} + \frac{h^2}{E} \left[\frac{1}{15} \left(12\Lambda(1+\nu) \overset{(1)}{S}{}^{\lambda}_{\alpha|\lambda} - \nu_* \overset{(1)}{S}{}^{\lambda}_{\lambda|\alpha} + \Lambda_* \overset{(1)}{\sigma}|_{\alpha} \right) \right. \\
& \left. - \frac{1}{35} \left(2\Lambda(1+\nu) \overset{(3)}{S}{}^{\lambda}_{\alpha|\lambda} - \nu_* \overset{(3)}{S}{}^{\lambda}_{\lambda|\alpha} + \Lambda_* \overset{(3)}{\sigma}|_{\alpha} \right) \right]
\end{aligned}
\tag{6.15a}
$$

$$
\begin{aligned}
\overset{(2)}{u}_{\alpha} = {} & -\frac{h^2}{E} \left[\frac{1}{3} \left(-\nu_* \overset{(0)}{S}{}^{\lambda}_{\lambda|\alpha} + \Lambda_* \overset{(0)}{\sigma}|_{\alpha} \right) - \frac{2}{21} \left(2\Lambda(1+\nu) \overset{(2)}{S}{}^{\lambda}_{\alpha|\lambda} - \nu_* \overset{(2)}{S}{}^{\lambda}_{\lambda|\alpha} + \Lambda_* \overset{(2)}{\sigma}|_{\alpha} \right) \right. \\
& \left. + \frac{1}{63} \left(2\Lambda(1+\nu) \overset{(4)}{S}{}^{\lambda}_{\alpha|\lambda} - \nu_* \overset{(4)}{S}{}^{\lambda}_{\lambda|\alpha} + \Lambda_* \overset{(4)}{\sigma}|_{\alpha} \right) \right]
\end{aligned}
\tag{6.15b}
$$

$$\overset{(n)}{u}_{\alpha} = -\frac{h^2}{E}\left[\frac{2\Lambda(1+\nu)\,\overset{(n-2)}{S}{}^{\lambda}{}_{\alpha|\lambda} - \nu_*\,\overset{(n-2)}{S}{}^{\lambda}{}_{\lambda|\alpha} + \Lambda_*\,\overset{(n-2)}{\sigma}|_{\alpha}}{(2n-3)(2n-1)} - 2\,\frac{2\Lambda(1+\nu)\,\overset{(n)}{S}{}^{\lambda}{}_{\alpha|\lambda} - \nu_*\,\overset{(n)}{S}{}^{\lambda}{}_{\lambda|\alpha} + \Lambda_*\,\overset{(n)}{\sigma}|_{\alpha}}{(2n-1)(2n+3)} + \frac{2\Lambda(1+\nu)\,\overset{(n+2)}{S}{}^{\lambda}{}_{\alpha|\lambda} - \nu_*\,\overset{(n+2)}{S}{}^{\lambda}{}_{\lambda|\alpha} + \Lambda_*\,\overset{(n+2)}{\sigma}|_{\alpha}}{(2n+3)(2n+5)}\right], \quad n \geq 3 \quad \textbf{(6.15c)}$$

The Legendre series for the laminar rotation follows from the introduction of the expansions (6.14) into formula (6.5b): noting the form of the displacement coefficients (6.15), we see that we may write

$$\psi = \sum_{n=0}^{\infty} \overset{(n)}{\psi}(x^{\lambda})\, P_n(t) \qquad \textbf{(6.16)}$$

in which the coefficients are given by

$$\overset{(0)}{\psi} = \frac{1}{2}\left(\overset{(0)}{u}_{1,2} - \overset{(0)}{u}_{2,1}\right) \qquad \textbf{(6.17a)}$$

$$\overset{(1)}{\psi} = \frac{h^2}{E}\Lambda(1+\nu)\left[\frac{2}{5}\left(\overset{(1)}{S}{}^{\lambda}{}_{1|\lambda 2} - \overset{(1)}{S}{}^{\lambda}{}_{2|\lambda 1}\right) - \frac{1}{35}\left(\overset{(3)}{S}{}^{\lambda}{}_{1|\lambda 2} - \overset{(3)}{S}{}^{\lambda}{}_{2|\lambda 1}\right)\right] \qquad \textbf{(6.17b)}$$

$$\overset{(2)}{\psi} = \frac{h^2}{E}\Lambda(1+\nu)\left[\frac{2}{21}\left(\overset{(2)}{S}{}^{\lambda}{}_{1|\lambda 2} - \overset{(2)}{S}{}^{\lambda}{}_{2|\lambda 1}\right) - \frac{1}{63}\left(\overset{(4)}{S}{}^{\lambda}{}_{1|\lambda 2} - \overset{(4)}{S}{}^{\lambda}{}_{2|\lambda 1}\right)\right] \qquad \textbf{(6.17c)}$$

$$\overset{(n)}{\psi} = -\frac{h^2}{E}\Lambda(1+\nu)\left[\frac{\overset{(n-1)}{S}{}^{\lambda}{}_{1|\lambda 2} - \overset{(n-1)}{S}{}^{\lambda}{}_{2|\lambda 1}}{(2n-3)(2n-1)} - 2\,\frac{\overset{(n)}{S}{}^{\lambda}{}_{1|\lambda 2} - \overset{(n)}{S}{}^{\lambda}{}_{2|\lambda 1}}{(2n-1)(2n+3)} + \frac{\overset{(n+2)}{S}{}^{\lambda}{}_{1|\lambda 2} - \overset{(n+2)}{S}{}^{\lambda}{}_{2|\lambda 1}}{(2n+3)(2n+5)}\right], \quad n \geq 3 \quad \textbf{(6.17d)}$$

For the undetermined quantities introduced by the integration of the constitutive relations, we note that

$$\frac{1}{2h}\int_{-h}^{h} v_\alpha\, dz = \frac{1}{2}\int_{-1}^{1} u_\alpha\, dt = \overset{(0)}{u}_\alpha,\quad \frac{1}{2h}\int_{-h}^{h} v_3\, dz = \frac{1}{2}\int_{-1}^{1} w\, dt = \overset{(0)}{w} \qquad \mathbf{(6.18)}$$

Hence, since $\overset{(0)}{u}_\alpha$ and $\overset{(0)}{w}$ measure the average through the plate thickness of the components of the displacement distribution, we refer to them as the mean laminar and mean transverse displacements, respectively. Similarly $\overset{(0)}{\psi}$ and $\overset{(0)}{\omega}_\alpha$, respectively, represent the mean values of the laminar and transverse components of infinitesimal rotation.

In the representations (6.7) and (6.14) for the displacement components, except for the undetermined mean values, all higher coefficients are given, by the systems of formulae (6.8) and (6.15), in terms of the elements appearing in the series for the stresses: in fact, recalling formulae (4.28) to (4.31) together with the general form (4.11), we see that all coefficients have, in fact, been determined in terms of the laminar stress coefficients of the original expansion (4.3). Thus all field quantities have been determined in terms of the unknown laminar stress coefficients and the undetermined mean displacements, the latter being the only displacement quantities to appear explicitly in the analysis.

7. The Equations for the Unknown Functions

The equations for the first two laminar stress coefficients have already been derived, namely equations (4.20) and (4.25). The system of equations for the higher laminar stress coefficients together with the equations for the determination of the mean displacements will follow from the third constitutive relation. If we introduce the representation (6.14) for the displacements together with the expansions (4.3) and (4.9) for the stress quantities into the constitutive relation (3.12c), the orthogonality of the Legendre polynomials requires that the resulting equation be satisfied term-by-term, yielding an infinite sequence of equations to be satisfied by the coefficients. Considering the equations in order, the first two in the sequence are for the determination of the mean displacements while the remainder constitute the system of equations to be satisfied by the higher laminar stress coefficients $\{\overset{(n)}{S}{}^{\alpha\beta}, n \geq 2\}$.

Utilizing the expansion (6.14) with the coefficients given by (6.15), the left hand side of (3.12c) may be put in the series form

$$\begin{aligned}\frac{1}{2}(u_{\alpha|\beta} + u_{\beta|\alpha}) &= \frac{1}{2}(\overset{(0)}{u}_{\alpha|\beta} + \overset{(0)}{u}_{\beta|\alpha})P_0 \\ &+ \Bigg\{ -h\,\overset{(0)}{w}|_{\alpha\beta} + \frac{h^2}{E}\Bigg[\frac{1}{15}\Big(6(1+\nu)(\overset{(1)}{S}{}^{\lambda}_{\alpha|\lambda\beta} + \overset{(1)}{S}{}^{\lambda}_{\beta|\lambda\alpha}) - \nu_*\overset{(1)}{S}{}^{\lambda}_{\lambda|\alpha\beta} + \Lambda_*\overset{(1)}{\sigma}|_{\alpha\beta}\Big) \\ &\qquad - \frac{1}{35}\Big(\Lambda(1+\nu)\,\overset{(3)}{S}{}^{\lambda}_{\alpha|\lambda\beta} + \overset{(3)}{S}{}^{\lambda}_{\beta|\lambda\alpha}) - \nu_*\overset{(3)}{S}{}^{\lambda}_{\lambda|\alpha\beta} + \Lambda_*\overset{(3)}{\sigma}|_{\alpha\beta}\Big)\Bigg]\Bigg\}P_1\end{aligned}$$

$$
\begin{aligned}
&-\frac{h^2}{E}\Bigg[\frac{1}{3}\Big(-\nu_*\overset{(0)}{S}{}^{\lambda}_{\lambda|\alpha\beta}+\Lambda_*\overset{(0)}{\sigma}_{|\alpha\beta}\Big)-\frac{2}{21}\Big(\Lambda(1+\nu)\big(\overset{(2)}{S}{}^{\lambda}_{\alpha|\lambda\beta}+\overset{(2)}{S}{}^{\lambda}_{\beta|\lambda\alpha}\big)-\nu_*\overset{(2)}{S}{}^{\lambda}_{\lambda|\alpha\beta}+\Lambda_*\overset{(2)}{\sigma}_{|\alpha\beta}\Big)\\
&\qquad+\frac{1}{63}\Big(\Lambda(1+\nu)\big(\overset{(4)}{S}{}^{\lambda}_{\alpha|\lambda\beta}+\overset{(4)}{S}{}^{\lambda}_{\beta|\lambda\alpha}\big)-\nu_*\overset{(4)}{S}{}^{\lambda}_{\lambda|\alpha\beta}+\Lambda_*\overset{(4)}{\sigma}_{|\alpha\beta}\Big)\Bigg]P_2\\
&-\frac{h^2}{E}\sum_{n=3}^{\infty}\Bigg[\frac{\Lambda(1+\nu)\big(\overset{(n-2)}{S}{}^{\lambda}_{\alpha|\lambda\beta}+\overset{(n-2)}{S}{}^{\lambda}_{\beta|\lambda\alpha}\big)-\nu_*\overset{(n-2)}{S}{}^{\lambda}_{\lambda|\alpha\beta}+\Lambda_*\overset{(n-2)}{\sigma}_{|\alpha\beta}}{(2n-3)(2n-1)}\\
&\qquad-2\,\frac{\Lambda(1+\nu)\big(\overset{(n)}{S}{}^{\lambda}_{\alpha|\lambda\beta}+\overset{(n)}{S}{}^{\lambda}_{\beta|\lambda\beta}\big)-\nu_*\overset{(n)}{S}{}^{\lambda}_{\lambda|\alpha\beta}+\Lambda_*\overset{(n)}{\sigma}_{|\alpha\beta}}{(2n-1)(2n+3)}\\
&\qquad+\frac{\Lambda(1+\nu)\big(\overset{(n+2)}{S}{}^{\lambda}_{\alpha|\lambda\beta}+\overset{(n+2)}{S}{}^{\lambda}_{\beta|\lambda\alpha}\big)-\nu_*\overset{(n+2)}{S}{}^{\lambda}_{\lambda|\alpha\beta}+\Lambda_*\overset{(n+2)}{\sigma}_{|\alpha\beta}}{(2n+3)(2n+5)}\Bigg]P_n
\end{aligned}
\tag{7.1}
$$

where we have used the fact that, in the plane, the order of covariant differentiation is immaterial. Similarly, employing the expansions (4.3) and (4.9), we obtain the Legendre series for the right hand side of (3.12c), namely

$$
\frac{(1+\nu)}{E}\sigma_{\alpha\beta}-g_{\alpha\beta}\frac{1}{E}[\nu\sigma^{\lambda}_{\lambda}+\nu_*\sigma]
=\frac{1}{E}\sum_{n=0}^{\infty}\Big[(1+\nu)\overset{(n)}{S}_{\alpha\beta}-g_{\alpha\beta}\big(\nu\overset{(n)}{S}{}^{\lambda}_{\lambda}+\nu_*\overset{(n)}{\sigma}\big)\Big]P_n \tag{7.2}
$$

Equating the series (7.1) with (7.2) yields the equations to be satisfied by the unknown functions.

The mean laminar displacements are to be determined from the equations resulting from the identification of the coefficients of P_0, namely

$$
\frac{1}{2}E\big(\overset{(0)}{u}_{\alpha|\beta}+\overset{(0)}{u}_{\beta|\alpha}\big)=(1+\nu)\overset{(0)}{S}_{\alpha\beta}-g_{\alpha\beta}\big(\nu\overset{(0)}{S}{}^{\lambda}_{\lambda}+\nu_*\overset{(0)}{\sigma}\big) \tag{7.3}
$$

while the corresponding identification of the coefficients of P_1 relates the mean transverse displacement to the stress coefficients in the form

$$
\begin{aligned}
-Eh\,\overset{(0)}{w}|_{\alpha\beta} = {} & (1+\nu)\,\overset{(1)}{S}{}_{\alpha\beta} - g_{\alpha\beta}\big(\nu\,\overset{(1)}{S}{}^{\lambda}_{\lambda} + \nu_*\,\overset{(1)}{\sigma}\big)\\
& - h^2\Bigg[\frac{1}{15}\Big(6\Lambda(1+\nu)\big(\overset{(1)}{S}{}^{\lambda}_{\alpha\,|\lambda\beta} + \overset{(1)}{S}{}^{\lambda}_{\beta\,|\lambda\alpha}\big) - \nu_*\,\overset{(1)}{S}{}^{\lambda}_{\lambda\,|\alpha\beta} + \Lambda_*\,\overset{(1)}{\sigma}|_{\alpha\beta}\Big)\\
& \qquad - \frac{1}{35}\Big(\Lambda(1+\nu)\big(\overset{(3)}{S}{}^{\lambda}_{\alpha\,|\lambda\beta} + \overset{(3)}{S}{}^{\lambda}_{\beta\,|\lambda\alpha}\big) - \nu_*\,\overset{(3)}{S}{}^{\lambda}_{\lambda\,|\alpha\beta} + \Lambda_*\,\overset{(3)}{\sigma}|_{\alpha\beta}\Big)\Bigg]
\end{aligned}
\tag{7.4}
$$

The system of equations satisfied by the higher laminar stress coefficients follows from the identification of the coefficients of the higher Legendre polynomials in the series (7.1) and (7.2): omitting the common factor $1/E$, we find

$$
\begin{aligned}
& (1+\nu)\,\overset{(2)}{S}{}_{\alpha\beta} - g_{\alpha\beta}\big(\nu\,\overset{(2)}{S}{}^{\lambda}_{\lambda} + \nu_*\,\overset{(2)}{\sigma}\big)\\
& + h^2\Bigg[\frac{1}{3}\big(-\nu_*\,\overset{(0)}{S}{}^{\lambda}_{\lambda\,|\alpha\beta} + \Lambda_*\,\overset{(0)}{\sigma}|_{\alpha\beta}\big) - \frac{2}{21}\Big(\Lambda(1+\nu)\big(\overset{(2)}{S}{}^{\lambda}_{\alpha\,|\lambda\beta} + \overset{(2)}{S}{}^{\lambda}_{\beta\,|\lambda\alpha}\big) - \nu_*\,\overset{(2)}{S}{}^{\lambda}_{\lambda\,|\alpha\beta} + \Lambda_*\,\overset{(2)}{\sigma}|_{\alpha\beta}\Big)\\
& \qquad + \frac{1}{63}\Big(\Lambda(1+\nu)\big(\overset{(4)}{S}{}^{\lambda}_{\alpha\,|\lambda\beta} + \overset{(4)}{S}{}^{\lambda}_{\beta\,|\lambda\alpha}\big) - \nu_*\,\overset{(4)}{S}{}^{\lambda}_{\lambda\,|\alpha\beta} + \Lambda_*\,\overset{(4)}{\sigma}|_{\alpha\beta}\Big)\Bigg] = 0
\end{aligned}
\tag{7.5a}
$$

$$
\begin{aligned}
& (1+\nu)\,\overset{(n)}{S}{}_{\alpha\beta} - g_{\alpha\beta}\big(\nu\,\overset{(n)}{S}{}^{\lambda}_{\alpha} + \nu_*\,\overset{(n)}{\sigma}\big)\\
& + h^2\Bigg[\frac{\Lambda(1+\nu)\big(\overset{(n-2)}{S}{}^{\lambda}_{\alpha\,|\lambda\beta} + \overset{(n-2)}{S}{}^{\lambda}_{\beta\,|\lambda\alpha}\big) - \nu_*\,\overset{(n-2)}{S}{}^{\lambda}_{\lambda\,|\alpha\beta} + \Lambda_*\,\overset{(n-2)}{\sigma}|_{\alpha\beta}}{(2n-3)(2n-1)}\\
& \qquad - 2\,\frac{\Lambda(1+\nu)\big(\overset{(n)}{S}{}^{\lambda}_{\alpha\,|\lambda\beta} + \overset{(n)}{S}{}^{\lambda}_{\beta\,|\lambda\alpha}\big) - \nu_*\,\overset{(n)}{S}{}^{\lambda}_{\lambda\,|\alpha\beta} + \Lambda_*\,\overset{(n)}{\sigma}|_{\alpha\beta}}{(2n-1)(2n+3)}\\
& \qquad + \frac{\Lambda(1+\nu)\big(\overset{(n+2)}{S}{}^{\lambda}_{\alpha\,|\lambda\beta} + \overset{(n+2)}{S}{}^{\lambda}_{\beta\,|\lambda\alpha}\big) - \nu_*\,\overset{(n+2)}{S}{}^{\lambda}_{\lambda\,|\alpha\beta} + \Lambda_*\,\overset{(n+2)}{\sigma}|_{\alpha\beta}}{(2n+3)(2n+5)}\Bigg] = 0\,,
\end{aligned}
$$

$$n \geq 3\,. \tag{7.5b}$$

We have now derived the full consequences of executing the integration with respect to the thickness coordinate. The system of equations consisting of (4.20), (4.25), (7.3), (7.4) and (7.5), when complemented by the edge conditions (5.14), constitute the two dimensional formulation of the boundary value problem previously posed in three dimensional terms by equations (3.10) to (3.12) with the boundary conditions (3.14) to (3.16). Prior to any further analysis of the above equations, we first observe some features of the problem brought into focus by the two dimensional formulation.

8. The Uncoupling of Effects: Principal and Residual Parts

An inspection of the two dimensional equations (4.20), (4.25), (7.3), (7.4) and (7.5) with the associated formulae for the transverse normal stress coefficients given by (4.11) and (4.12) and for the transverse shear stress coefficients by (4.21), (4.26), (4.27) and the general form (4.7), together with the edge conditions (5.14) shows that the general problem uncouples to yield the two distinct problems of stretching and bending characterized as follows

1. The problem of stretching is concerned with the quantities with even index n in the sequences $\{\overset{(n)}{S}{}^{\alpha\beta}\}$ and $\{\overset{(n)}{\sigma}\}$ together with the mean laminar displacements $\overset{(0)}{u}_\alpha$: also involved are the elements with even index n in the sequences $\{\overset{(n)}{u}_\alpha\}$ and $\{\overset{(n)}{\psi}\}$ and the quantities with odd index n in the sequences $\{\overset{(n)}{\tau}{}^\alpha\}$, $\{\overset{(n)}{w}\}$ and $\{\overset{(n)}{\omega}\}$.

2. The problem of bending is concerned with the quantities with odd index n in the sequences $\{\overset{(n)}{S}{}^{\alpha\beta}\}$ and $\{\overset{(n)}{\sigma}\}$ together with the mean transverse displacement $\overset{(0)}{w}$ and mean transverse rotations $\overset{(0)}{\omega}{}^\alpha$: also involved are the elements with even index n in the sequences $\{\overset{(n)}{\tau}{}^\alpha\}$, $\{\overset{(n)}{w}\}$ and $\{\overset{(n)}{\omega}_\alpha\}$ together with the quantities with odd index n in the sequences $\{\overset{(n)}{u}_\alpha\}$ and $\{\overset{(n)}{\psi}\}$.

In particular equations (7.3) is associated with equations (4.20) while equations (7.4) bear a corresponding relation to equation (4.25).

In the sequence $\{\overset{(n)}{S}{}^{\alpha\beta}\}$ the coefficients $\overset{(0)}{S}{}^{\alpha\beta}$ and $\overset{(1)}{S}{}^{\alpha\beta}$, respectively measuring the laminar stress resultants and stress couples, are the principal laminar stress effects; the higher coefficients $\{\overset{(n)}{S}{}^{\alpha\beta}, n \geq 2\}$, having no effect on either the stress resultants or stress couples, are termed the residual laminar stress effects. The components of the principal stretching coefficients tensor $\overset{(0)}{S}{}^{\alpha\beta}$ must satisfy the system of equations (4.20) while equation (4.25) furnishes the corresponding equation satisfied by the principal bending coefficient tensor $\overset{(1)}{S}{}^{\alpha\beta}$. However, neither the system (4.20) nor equation (4.25) is determinate so that neither can be solved without reference to the equations for the displacement quantities.

In addition to the principal elements, equations (7.3) and (7.4) for the mean displacements also involve some of the residual stress coefficients. Inspection of these equations, therefore, clarifies how the principal and residual components of the mean displacements are to be defined so as to yield an uncoupled formulation of the equations for the principal effects in both the stretching and bending problems. A necessary preliminary to this uncoupling is the appropriate resolution of the transverse normal stress into the principal and residual parts: this is most conveniently done in the context of the linear vector space induced by the Legendre expansion of the field quantities.

There, then, remains the residual problems involving the equations for the residual mean displacements combined with equations (7.5) for the residual stress coefficients. These latter still involve some of the principal stress coefficients so that the analysis of the residual problem presupposes the prior solution of the equations for the principal effects. These principal stress coefficients then appear as inhomogeneous terms in the equations for the residual stresses.

The manner of effecting this general resolution, leading to the partial uncoupling of the principal and residual problems, is best illustrated by considering the stretching and bending problems individually.

9. The Problem of Stretching

In the problem of stretching, the surface forces are applied symmetrically, that is

$$\overset{+}{p} = \overset{-}{p} = \frac{1}{2}p^*, \; p = 0 \tag{9.1}$$

while the specification of the stress vectors on the edge surface is such that, in their dependence on the thickness coordinate, the laminar components are even while the transverse shear components are odd: hence,

$$\overset{(2k+1)}{T^{\alpha}}(\eta) = 0, \qquad \overset{(2k+1)}{D}(\eta) = 0, \; k \geq 0 \,. \tag{9.2}$$

The Legendre representations for the stress components, therefore, take the form

$$\sigma^{\alpha\beta} = \sum_{k=0}^{\infty} \overset{(2k)}{S^{\alpha\beta}}(x^{\lambda})\, P_{2k}(t) \tag{9.3a}$$

$$\tau^{\alpha} = \sum_{k=0}^{\infty} \overset{(2k+1)}{\tau^{\alpha}}(x^{\lambda})\, P_{2k+1}(t) \tag{9.3b}$$

$$\sigma = \sum_{k=0}^{\infty} \overset{(2k)}{\sigma}(x^{\lambda})\, P_{2k}(t) \tag{9.3c}$$

in which the zero-th coefficients satisfy the pair of equations

$$\overset{(0)}{S}{}^{\beta\alpha}|_\beta = 0 \tag{9.4}$$

$$\overset{(0)}{\sigma} = \frac{1}{3} h \overset{(1)}{\tau}{}^{\alpha}|_\alpha + \frac{1}{2} p^* \tag{9.5}$$

reproduced respectively from (4.20) and (4.23). Equations (9.4) are clearly for the determination of the principal laminar stretching coefficients.

We shall presently formulate the equations for the higher laminar stress coefficients in the expansion (9.3a) in terms of which, recalling (4.26) and the general form (4.7), the shear stress coefficients in the expansion (9.3b) are given by

$$\overset{(1)}{\tau}{}^{\alpha} = \frac{1}{5} h \overset{(2)}{S}{}^{\beta\alpha}|_\beta \tag{9.6a}$$

$$\overset{(2k+1)}{\tau}{}^{\alpha} = h \left[\frac{\overset{(2k+2)}{S}{}^{\beta\alpha}|_\beta}{4k+5} - \frac{\overset{(2k)}{S}{}^{\beta\alpha}|_\beta}{4k+1} \right], \quad k \geq 1 \tag{9.6b}$$

while, from (4.28), (4.30) and the general form (4.11), the corresponding formulae for the transverse normal stress coefficients appearing in (9.3c) have the form

$$\overset{(0)}{\sigma} = \frac{1}{15} h^2 \overset{(2)}{S}{}^{\lambda}{}_{\mu}|^{\mu}_{\lambda} + \frac{1}{2} p^* \tag{9.7a}$$

$$\overset{(2)}{\sigma} = h^2 \left[-\frac{2}{21} \overset{(2)}{S}{}^{\lambda}{}_{\mu}|^{\mu}_{\lambda} + \frac{1}{63} \overset{(4)}{S}{}^{\lambda}{}_{\mu}|^{\mu}_{\lambda} \right] \tag{9.7b}$$

$$\overset{(2k)}{\sigma} = h^2 \left[\frac{\overset{(2k-2)}{S}{}^{\lambda}{}_{\mu}|^{\mu}_{\lambda}}{(4k-3)(4k-1)} - \frac{2\, \overset{(2k)}{S}{}^{\lambda}{}_{\mu}|^{\mu}_{\lambda}}{(4k-1)(4k+3)} + \frac{\overset{(2k+2)}{S}{}^{\lambda}{}_{\mu}|^{\mu}_{\lambda}}{(4k+3)(4k+5)} \right], \quad k \geq 2 \tag{9.7c}$$

where, in the invariant quantities on the right, we have altered the indexing to a form more convenient for later use.

Corresponding to the representations (9.3) the Legendre series for the displacement components are written

$$u_\alpha = \sum_{k=0}^{\infty} \overset{(2k)}{u}_\alpha(x^\lambda)\, P_{2k}(t) \tag{9.8a}$$

$$w = \sum_{k=0}^{\infty} \overset{(2k+1)}{w}(x^{\lambda})\, P_{2k}(t) \tag{9.8b}$$

in which only the $\overset{(0)}{u}_{\alpha}$ remain to be determined, since, with the transverse normal stress coefficients given by (9.7), relations (6.15b,c) render the higher laminar displacement coefficients in terms of the laminar stress coefficients according to the formulae

$$\overset{(2)}{u}_{\alpha} = -\frac{h^2}{E}\Bigg[\frac{1}{3}\Big(-\nu_*\, \overset{(0)}{S}{}^{\lambda}{}_{\lambda|\alpha} + \Lambda_*\, \overset{(0)}{\sigma}|_{\alpha}\Big) - \frac{2}{21}\Big(2\Lambda(1+\nu)\, \overset{(2)}{S}{}^{\lambda}{}_{\alpha|\lambda} - \nu_*\, \overset{(2)}{S}{}^{\lambda}{}_{\lambda|\alpha} + \Lambda_*\, \overset{(2)}{\sigma}|_{\alpha}\Big)$$
$$+ \frac{1}{63}\Big(2\Lambda(1+\nu)\, \overset{(4)}{S}{}^{\lambda}{}_{\alpha|\lambda} - \nu_*\, \overset{(4)}{S}{}^{\lambda}{}_{\lambda|\alpha} + \Lambda_*\, \overset{(4)}{\sigma}|_{\alpha}\Big)\Bigg] \tag{9.9a}$$

$$\overset{(2k)}{u}_{\alpha} = -\frac{h^2}{E}\Bigg[\frac{2\Lambda(1+\nu)\, \overset{(2k-2)}{S}{}^{\lambda}{}_{\alpha|\lambda} - \nu_*\, \overset{(2k-2)}{S}{}^{\lambda}{}_{\lambda|\alpha} + \Lambda_*\, \overset{(2k-2)}{\sigma}|_{\alpha}}{(4k-3)(4k-1)}$$
$$-2\,\frac{2\Lambda(1+\nu)\, \overset{(2k)}{S}{}^{\lambda}{}_{\alpha|\lambda} - \nu_*\, \overset{(2k)}{S}{}^{\lambda}{}_{\lambda|\alpha} + \Lambda_*\, \overset{(2k)}{\sigma}|_{\alpha}}{(4k-1)(4k+3)}$$
$$+\frac{2\Lambda(1+\nu)\, \overset{(2k+2)}{S}{}^{\lambda}{}_{\alpha|\lambda} - \nu_*\, \overset{(2k+2)}{S}{}^{\lambda}{}_{\lambda|\alpha} + \Lambda_*\, \overset{(2k+2)}{\sigma}|_{\alpha}}{(4k+3)(4k+5)}\Bigg], \quad k \geq 2 \tag{9.9b}$$

while for the transverse displacement coefficients in the series (9.8b), we have the corresponding form

$$\overset{(2k+1)}{w} = -\frac{h}{E}\Bigg[\frac{\Lambda_*\, \overset{(2k+2)}{\sigma} - \nu_*\, \overset{(2k+2)}{S}{}^{\lambda}{}_{\lambda}}{4k+5} - \frac{\Lambda_*\, \overset{(2k)}{\sigma} - \nu_*\, \overset{(2k)}{S}{}^{\lambda}{}_{\lambda}}{4k+1}\Bigg], \quad k \geq 0 \tag{9.10}$$

as follows from the general relation (6.8).

For the rotation components, we have the expansions

$$\psi = \sum_{k=0}^{\infty} \overset{(2k)}{\psi}(x^{\lambda})\, P_{2k}(t) \tag{9.11a}$$

$$\omega_{\alpha} = \sum_{k=0}^{\infty} \overset{(2k+1)}{\omega}_{\alpha}(x^{\lambda})\, P_{2k+1}(t) \tag{9.11b}$$

in which, from (6.17), the laminar elements are given by

$$\overset{(0)}{\psi} = \frac{1}{2}(\overset{(0)}{u}_{1,2} - \overset{(0)}{u}_{2,1}) \tag{9.12a}$$

$$\overset{(2)}{\psi} = \frac{h^2}{E}\Lambda(1+\nu)\left[\frac{2}{21}(\overset{(2)}{S}{}^{\lambda}_{1|\lambda 2} - \overset{(2)}{S}{}^{\lambda}_{2|\lambda 1}) - \frac{1}{63}(\overset{(4)}{S}{}^{\lambda}_{1|\lambda 2} - \overset{(4)}{S}{}^{\lambda}_{2|\lambda 1})\right] \tag{9.12b}$$

$$\overset{(2k)}{\psi} = -\frac{h^2}{E}\Lambda(1+\nu)\left[\frac{\overset{(2k-2)}{S}{}^{\lambda}_{1|\lambda 2} - \overset{(2k-2)}{S}{}^{\lambda}_{2|\lambda 1}}{(4k-3)(4k-1)} - 2\frac{\overset{(2k)}{S}{}^{\lambda}_{1|\lambda 2} - \overset{(2k)}{S}{}^{\lambda}_{2|\lambda 1}}{(4k-1)(4k+3)} + \frac{\overset{(2k+2)}{S}{}^{\lambda}_{1|\lambda 2} - \overset{(2k+2)}{S}{}^{\lambda}_{2|\lambda 1}}{(4k+3)(4k+5)}\right], \quad k \geq 2 \tag{9.12c}$$

and the coefficients in the expansion (9.11b), for the components of transverse rotation, are

$$\overset{(1)}{\omega}_{\alpha} = \frac{h}{E}\left[\frac{1}{5}\left(\Lambda(1+\nu)\overset{(2)}{S}{}^{\lambda}_{\alpha|\lambda} - \nu_*\overset{(2)}{S}{}^{\lambda}_{\lambda|\alpha} + \Lambda_*\overset{(2)}{\sigma}|_{\alpha}\right) - \left(-\nu_*\overset{(0)}{S}{}^{\lambda}_{\lambda|\alpha} + \Lambda_*\overset{(0)}{\sigma}|_{\alpha}\right)\right] \tag{9.13a}$$

$$\overset{(2k+1)}{\omega}_{\alpha} = \frac{h}{E}\left[\frac{\Lambda(1+\nu)\overset{(2k+2)}{S}{}^{\lambda}_{\alpha|\lambda} - \nu_*\overset{(2k+2)}{S}{}^{\lambda}_{\lambda|\alpha} + \Lambda_*\overset{(2k+2)}{\sigma}|_{\alpha}}{4k+5} - \frac{\Lambda(1+\nu)\overset{(2k)}{S}{}^{\lambda}_{\alpha|\lambda} - \nu_*\overset{(2k)}{S}{}^{\lambda}_{\lambda|\alpha} + \Lambda_*\overset{(2k)}{\sigma}|_{\alpha}}{4k+1}\right], \quad k \geq 1 \tag{9.13b}$$

as derived in (6.12).

Equations (7.3) for the determination of the mean laminar displacements, on the introduction of formula (9.7a) for $\overset{(0)}{\sigma}$, take the form

$$\frac{1}{2}E(\overset{(0)}{u}_{\alpha|\beta} + \overset{(0)}{u}_{\beta|\alpha}) = (1+\nu)\overset{(0)}{S}_{\alpha\beta} - g_{\alpha\beta}[\nu\overset{(0)}{S}{}^{\lambda}_{\lambda} + \nu_*(\frac{1}{2}p^* + \frac{1}{15}h^2\overset{(2)}{S}{}^{\lambda\mu}_{\mu|\lambda})] \tag{9.14}$$

From (7.5) with even index, we have the system of equations for the residual laminar stress coefficients: if, in (7.5a), we again use formula (9.7a) for $\overset{(0)}{\sigma}$, then the system of equations may be written

$$
\begin{aligned}
&(1+\nu)\,\overset{(2)}{S}_{\alpha\beta} - g_{\alpha\beta}\Big(\nu\,\overset{(2)}{S}{}^{\lambda}_{\lambda} + \nu_*\,\overset{(2)}{\sigma}\Big) + \frac{1}{45}\Lambda_* h^4\,\overset{(2)}{S}{}^{\lambda}{}_{\mu}|^{\mu}{}_{\lambda\alpha\beta} \\
&+ h^2\Bigg[-\frac{2}{21}\Big(\Lambda(1+\nu)\big(\overset{(2)}{S}{}^{\lambda}{}_{\alpha|\lambda\beta} + \overset{(2)}{S}{}^{\lambda}{}_{\beta|\lambda\alpha}\big) - \nu_*\,\overset{(2)}{S}{}^{\lambda}{}_{\lambda|\alpha\beta} + \Lambda_*\,\overset{(2)}{\sigma}|_{\alpha\beta}\Big) \\
&\qquad + \frac{1}{63}\Big(\Lambda(1+\nu)\big(\overset{(4)}{S}{}^{\lambda}{}_{\alpha|\lambda\beta} + \overset{(4)}{S}{}^{\lambda}{}_{\beta|\lambda\alpha}\big) - \nu_*\,\overset{(4)}{S}{}^{\lambda}{}_{\lambda|\alpha\beta} + \Lambda_*\,\overset{(4)}{\sigma}|_{\alpha\beta}\Big)\Bigg] \\
&\qquad\qquad = \frac{1}{3}\nu_* h^2\,\overset{(0)}{S}{}^{\lambda}{}_{\lambda|\alpha\beta} - \frac{1}{6}\Lambda_* h^2 p^*|_{\alpha\beta}
\end{aligned}
\tag{9.15a}
$$

$$
\begin{aligned}
&(1+\nu)\,\overset{(2k)}{S}_{\alpha\beta} - g_{\alpha\beta}\Big(\nu\,\overset{(2k)}{S}{}^{\lambda}_{\lambda} + \nu_*\,\overset{(2k)}{\sigma}\Big) \\
&+ h^2\Bigg[\frac{\Lambda(1+\nu)\big(\overset{(2k-2)}{S}{}^{\lambda}{}_{\alpha|\lambda\beta} + \overset{(2k-2)}{S}{}^{\lambda}{}_{\beta|\lambda\alpha}\big) - \nu_*\,\overset{(2k-2)}{S}{}^{\lambda}{}_{\lambda|\alpha\beta} + \Lambda_*\,\overset{(2k-2)}{\sigma}|_{\alpha\beta}}{(4k-3)(4k-1)} \\
&\qquad - 2\,\frac{\Lambda(1+\nu)\big(\overset{(2k)}{S}{}^{\lambda}{}_{\alpha|\lambda\beta} + \overset{(2k)}{S}{}^{\lambda}{}_{\beta|\lambda\alpha}\big) - \nu_*\,\overset{(2k)}{S}{}^{\lambda}{}_{\lambda|\alpha\beta} + \Lambda_*\,\overset{(2k)}{\sigma}|_{\alpha\beta}}{(4k-1)(4k+3)} \\
&\qquad + \frac{\Lambda(1+\nu)\big(\overset{(2k+2)}{S}{}^{\lambda}{}_{\alpha|\lambda\beta} + \overset{(2k+2)}{S}{}^{\lambda}{}_{\beta|\lambda\alpha}\big) - \nu_*\,\overset{(2k+2)}{S}{}^{\lambda}{}_{\lambda|\alpha\beta} + \Lambda_*\,\overset{(2k+2)}{\sigma}|_{\alpha\beta}}{(4k+3)(4k+5)}\Bigg] = 0,\ k \geq 2
\end{aligned}
\tag{9.15b}
$$

For each of the field quantities, the expansion in terms of Legendre functions yields a component representation in an infinite dimensional linear space. Defining the residual effects as those induced exclusively by the residual laminar stress coefficients, the resolution of the stress and displacement fields into their principal and residual parts follows from an inspection of these component representations. Thus, for the transverse normal stress, we have

$$
\sigma = \{\,\overset{(2k)}{\sigma},\ 0 \leq k \leq \infty\}
\tag{9.16}
$$

in which each component may be considered the sum of a principal effect $\overset{(2k)}{\sigma}{}^{P}$ and a residual effect $\overset{(2k)}{\sigma}{}^{R}$ so that

$$\overset{(2k)}{\sigma} = \overset{(2k)}{\sigma}{}^{P} + \overset{(2k)}{\sigma}{}^{R} \tag{9.17}$$

From (9.7) we see that

$$\overset{(0)}{\sigma}{}^{P} = \frac{1}{2}p^*, \quad \overset{(0)}{\sigma}{}^{R} = \frac{1}{15}h^2 \overset{(2)}{S}{}^{\lambda}{}_{\mu}{}^{|\mu}{}_{|\lambda} \tag{9.18a,b}$$

$$\overset{(2k)}{\sigma}{}^{P} = 0, \quad \overset{(2k)}{\sigma}{}^{R} = \overset{(2k)}{\sigma}, \; k \geq 1 \tag{9.18c,d}$$

Hence, if we write

$$\sigma^{P} = \{ \overset{(2k)}{\sigma}{}^{P}, 0 \leq k \leq \infty \}, \sigma^{R} = \{ \overset{(2k)}{\sigma}{}^{R}, 0 \leq k \leq \infty \} \tag{9.19}$$

then, with the standard convention for vector summation, we have

$$\sigma = \sigma^{P} + \sigma^{R} \tag{9.20}$$

The corresponding resolution of the other quantities follows the same pattern.

Returning now to equation (9.14), if we let $\overset{(R)}{U}_{\alpha}$ denote the residual components of the mean laminar displacements, then for their determination, we have the system of equations

$$\frac{1}{2}E(\overset{(R)}{U}_{\alpha|\beta} + \overset{(R)}{U}_{\beta|\alpha}) = -g_{\alpha\beta}\frac{1}{15}\nu_{*}h^2 \overset{(2)}{S}{}^{\lambda}{}_{\mu}{}^{|\mu}{}_{|\lambda} \tag{9.21}$$

Presuming that the principal quantities $\overset{(0)}{S}{}^{\lambda}{}_{\lambda|\alpha\beta}$ have already been calculated, the set of equations (9.15), combined with equations (9.21), when complemented by the edge conditions (5.14c) with even index n, namely

$$\overset{(2k)}{S}{}^{\beta\alpha} n_{\beta} = \overset{(2k)}{T}{}^{\alpha}, \; h \overset{(2k)}{S}{}^{\beta\alpha}{}_{|\beta} n_{\alpha} = \overset{(2k)}{D}, \; k \geq 1 \tag{9.22}$$

constitute the boundary value problem for the residual effects.

Writing U_{α} for the principal components of the mean laminar displacements so that

$$\overset{(0)}{U}_{\alpha} = U_{\alpha} + \overset{(R)}{U}_{\alpha} \tag{9.23}$$

then by subtracting equation (9.21) from (9.14), we have

$$\frac{1}{2}E(U_{\alpha|\beta} + U_{\beta|\alpha}) = (1+\nu)\overset{(0)}{S}_{\alpha\beta} - g_{\alpha\beta}(\nu \overset{(0)}{S}{}^{\lambda}_{\lambda} + \frac{1}{2}\nu_{*}p^*) \tag{9.24}$$

which, when combined with the system of equations (9.4) and the edge conditions (5.14a), namely

$$\overset{(0)}{S}{}^{\beta\alpha} n_\beta = \overset{(0)}{T}{}^{\alpha} \tag{9.25}$$

constitute the boundary value problem for the principal effects.

9P. The Principal Stretching Problem: (Generalized Plane Stress)

The principal stretching problem, described by equations (9.4) and (9.24) with conditions (9.25) is the classical problem of generalized plane stress. We follow the standard method of reduction by observing that equations (9.4) are identically satisfied if we represent the stress components in the form

$$\overset{(0)}{S}{}^{\alpha\beta} = \epsilon^{\alpha\lambda} \epsilon^{\beta\mu} F|_{\lambda\mu} \tag{9P.1}$$

where F is the Airy stress function to be determined from a compatibility equation implied by the constitutive relations (9.24).

The trace of the system of tensor equations (9.24) yields the relation

$$EU^\alpha|_\alpha = (1-\nu)\, \overset{(0)}{S}{}^{\alpha}_{\alpha} - \nu_* p^* \tag{9P.2}$$

We next consider the contravariant form of (9.24) and take a repeated covariant derivative, first with respect to the variable x^α and then with respect to the variable x^β: after summing over the repeated indices, we utilize equations (9.4) and find

$$EU^\alpha|^\beta_{\alpha\beta} = -\nu\, \overset{(0)}{S}{}^{\lambda}_{\lambda}|^{\alpha}_{\alpha} - \frac{1}{2}\nu_* p^*|^\alpha_\alpha \tag{9P.3}$$

By applying the Laplacian operator to equation (9P.2), we obtain a second form of the above relation, namely

$$EU^\alpha|^\beta_{\alpha\beta} = (1-\nu)\, \overset{(0)}{S}{}^{\alpha}_{\alpha}|^{\beta}_{\beta} - \nu_* p^*|^\beta_\beta \tag{9P.4}$$

The consistency of the two forms (9P.3) and (9P.4) requires the satisfaction of the compatibility condition

$$\overset{(0)}{S}{}^{\alpha}_{\alpha}|^{\beta}_{\beta} = \frac{1}{2}\nu_* p^*|^\alpha_\alpha \tag{9P.5}$$

Recalling relation (1.12), it follows from (9P.1) that

$$\overset{(0)}{S}{}^{\alpha}_{\alpha} = F|^{\alpha}_{\alpha} \tag{9P.6}$$

so that in terms of the stress function equation (9P.5) reads

$$F|^{\alpha\beta}_{\alpha\beta} = \frac{1}{2}\nu_* p^*|^{\alpha}_{\alpha} \tag{9P.7}$$

which, in the absence of surface pressure becomes the well known biharmonic equation for the determination of the Airy function. In terms of the latter, the associated edge conditions (9.25) read

$$\epsilon^{\beta\lambda}\epsilon^{\alpha\mu}F|_{\lambda\mu}\,n_{\beta} = \overset{(0)}{T}{}^{\alpha} \tag{9P.8}$$

complementing the above equation.

9R. The Residual Stretching Problem

In the further reduction of the equations for the residual problem, we develop the treatment along the lines followed for the equations for the principal effects. Taking the trace of the tensor equation (9.21), we have

$$E\,\overset{(R)}{U}{}^{\alpha}|_{\alpha} = -\frac{2}{15}\nu_*\,\overset{(2)}{S}{}^{\lambda}_{\mu}|^{\mu}_{\lambda} \tag{9R.1}$$

Next we consider the contravariant form of (9.21) and, having taken the repeated covariant derivative with respect to the variables x^{α} and x^{β}, we sum over the repeated indices to obtain

$$E\,\overset{(R)}{U}{}^{\alpha}|^{\beta}_{\alpha\beta} = -\frac{1}{15}\nu_*\,\overset{(2)}{S}{}^{\lambda}_{\mu}|^{\mu\alpha}_{\lambda\alpha} \tag{9R.2}$$

For the consistency of relations (9R.1) and (9R.2), we must have

$$\overset{(2)}{S}{}^{\lambda}_{\mu}|^{\mu\alpha}_{\lambda\alpha} = 0 \tag{9R.3}$$

and hence also

$$\overset{(R)}{U}{}^{\alpha}|^{\beta}_{\alpha\beta} = 0 \tag{9R.4}$$

Turning to equations (9.15), we derive the relations resulting from taking the trace of each tensor equation, namely

$$
\begin{aligned}
&(1-\nu)\,\overset{(2)}{S}{}^{\alpha}_{\alpha} - 2\nu_*\,\overset{(2)}{\sigma} \\
&\quad + h^2\Bigg[-\frac{2}{21}\Big(2\Lambda(1+\nu)\,\overset{(2)}{S}{}^{\lambda}_{\alpha}\big|^{\alpha}_{\lambda} - \nu_*\,\overset{(2)}{S}{}^{\lambda}_{\lambda}\big|^{\alpha}_{\alpha} + \Lambda_*\,\overset{(2)}{\sigma}\big|^{\alpha}_{\alpha}\Big) \\
&\qquad + \frac{1}{63}\Big(2\Lambda(1+\nu)\,\overset{(4)}{S}{}^{\lambda}_{\alpha}\big|^{\alpha}_{\lambda} - \nu_*\,\overset{(4)}{S}{}^{\lambda}_{\lambda}\big|^{\alpha}_{\alpha} + \Lambda_*\,\overset{(4)}{\sigma}\big|^{\alpha}_{\alpha}\Big)\Bigg] = \frac{1}{6}(\nu_*^2 - \Lambda_*)h^2 p^*\big|^{\alpha}_{\alpha}
\end{aligned}
\tag{9R.5a}
$$

$$
\begin{aligned}
&(1-\nu)\,\overset{(2k)}{S}{}^{\alpha}_{\alpha} - 2\nu_*\,\overset{(2k)}{\sigma} \\
&\quad + h^2\Bigg[\frac{2\Lambda(1+\nu)\,\overset{(2k-2)}{S}{}^{\lambda}_{\alpha}\big|^{\alpha}_{\lambda} - \nu_*\,\overset{(2k-2)}{S}{}^{\lambda}_{\lambda}\big|^{\alpha}_{\alpha} + \Lambda_*\,\overset{(2k-2)}{\sigma}\big|^{\alpha}_{\alpha}}{(4k-3)(4k-1)} \\
&\qquad - 2\,\frac{2\Lambda(1+\nu)\,\overset{(2k)}{S}{}^{\lambda}_{\alpha}\big|^{\alpha}_{\lambda} - \nu_*\,\overset{(2k)}{S}{}^{\lambda}_{\lambda}\big|^{\alpha}_{\alpha} + \Lambda_*\,\overset{(2k)}{\sigma}\big|^{\alpha}_{\alpha}}{(4k-1)(4k+3)} \\
&\qquad + \frac{2\Lambda(1+\nu)\,\overset{(2k+2)}{S}{}^{\lambda}_{\alpha}\big|^{\alpha}_{\lambda} - \nu_*\,\overset{(2k+2)}{S}{}^{\lambda}_{\lambda}\big|^{\alpha}_{\alpha} + \Lambda_*\,\overset{(2k+2)}{\sigma}\big|^{\alpha}_{\alpha}}{(4k+3)(4k+5)}\Bigg] = 0,\ k \geq 0
\end{aligned}
\tag{9R.5b}
$$

where in (9R.5a) we have noted relation (9R.3) and have substituted from (9P.5) on the right.

We next consider the contravariant form of (9.15a) and taking the repeated covariant derivative with respect to the variables x^{α} and x^{β}, we sum over the repeated indices: if we again substitute from (9R.3) and (9P.5), we find

$$
\begin{aligned}
&(1+\nu)\,\overset{(2)}{S}{}^{\alpha}_{\beta}\big|^{\beta}_{\alpha} - \nu\,\overset{(2)}{S}{}^{\lambda}_{\lambda}\big|^{\alpha}_{\alpha} - \nu_*\,\overset{(2)}{\sigma}\big|^{\alpha}_{\alpha} \\
&\quad + h^2\Bigg[-\frac{2}{21}\Big(-\nu_*\,\overset{(2)}{S}{}^{\lambda}_{\lambda}\big|^{\alpha\beta}_{\alpha\beta} + \Lambda_*\,\overset{(2)}{\sigma}\big|^{\alpha\beta}_{\alpha\beta}\Big) + \frac{1}{63}\Big(2\Lambda\,\overset{(4)}{S}{}^{\lambda}_{\alpha}\big|^{\alpha\beta}_{\lambda\beta} - \nu_*\,\overset{(4)}{S}{}^{\lambda}_{\lambda}\big|^{\alpha\beta}_{\alpha\beta} + \Lambda_*\,\overset{(4)}{\sigma}\big|^{\alpha\beta}_{\alpha\beta}\Big)\Bigg] \\
&\qquad = \frac{1}{6}(\nu_*^2 - \Lambda_*)h^2 p^*\big|^{\alpha\beta}_{\alpha\beta}
\end{aligned}
\tag{9R.6a}
$$

A similar operation on each of the tensor equations (9.15b) yields

$$
(1+\nu)\ \overset{(2k)}{S}{}^{\alpha}_{\beta}|^{\beta}_{\alpha} - \nu\ \overset{(2k)}{S}{}^{\lambda}_{\lambda}|^{\alpha}_{\alpha} - \nu_*\ \overset{(2k)}{\sigma}|^{\alpha}_{\alpha}
$$

$$
+ h^2 \left[\frac{2\Lambda(1+\nu)\ \overset{(2k-2)}{S}{}^{\lambda}_{\alpha}|^{\alpha\beta}_{\lambda\beta} - \nu_*\ \overset{(2k-2)}{S}{}^{\lambda}_{\lambda}|^{\alpha\beta}_{\alpha\beta} + \Lambda_*\ \overset{(2k-2)}{\sigma}|^{\alpha\beta}_{\alpha\beta}}{(4k-3)(4k-1)} \right.
$$

$$
- 2\, \frac{2\Lambda(1+\nu)\ \overset{(2k)}{S}{}^{\lambda}_{\alpha}|^{\alpha\beta}_{\lambda\beta} - \nu_*\ \overset{(2k)}{S}{}^{\lambda}_{\lambda}|^{\alpha\beta}_{\alpha\beta} + \Lambda_*\ \overset{(2k)}{\sigma}|^{\alpha\beta}_{\alpha\beta}}{(4k-1)(4k+3)}
$$

$$
\left. + \frac{2\Lambda(1+\nu)\ \overset{(2k+2)}{S}{}^{\lambda}_{\alpha}|^{\alpha\beta}_{\lambda\beta} - \nu_*\ \overset{(2k+2)}{S}{}^{\lambda}_{\lambda}|^{\alpha\beta}_{\alpha\beta}) + \Lambda_*\ \overset{(2k+2)}{\sigma}|^{\alpha\beta}_{\alpha\beta}}{(4k+3)(4k+5)} \right] = 0,\ k \geq 2 \quad \textbf{(9R.6b)}
$$

The requirement that relations (9R.6) be consistent respectively with the associated relations in (9R.5) leads to a set of compatibility conditions. We first apply the Laplacian operator to equations (9R.5) and utilize relation (9R.3) in the formula resulting from (9R.5a): if we then subtract these derived relations respectively from the corresponding equations in (9R.6), we obtain the sequence of relations

$$
(1+\nu)\ \overset{(2k)}{S}{}^{\alpha}_{\beta}|^{\beta}_{\alpha} - \overset{(2k)}{S}{}^{\lambda}_{\lambda}|^{\alpha}_{\alpha} + \nu_*\ \overset{(2k)}{\sigma}|^{\alpha}_{\alpha} = 0,\ k \geq 1 \qquad \textbf{(9R.7)}
$$

With the $\overset{(2k)}{\sigma}$, $k \geq 1$, given by (9.7b,c), and also noting (9R.3), we see that the sets (9R.5) and (9R.7) constitute two systems of equations for the simultaneous determination of the two sequences of invariants $\overset{(2k)}{S}{}^{\lambda}_{\lambda}$, $k \geq 1$ and $\overset{(2k)}{S}{}^{\lambda}_{\mu}|^{\mu}_{\lambda}$ $k \geq 1$.

The equations for the determination of the individual components will follow from a little modification of equations (9.15). For this we shall use the identities

$$
\overset{(2k)}{S}{}^{\lambda}_{\alpha|\lambda\beta} + \overset{(2k)}{S}{}^{\lambda}_{\beta|\lambda\alpha} = \overset{(2k)}{S}{}_{\alpha\beta|\lambda}{}^{\lambda} + \overset{(2k)}{S}{}^{\lambda}_{\lambda|\alpha\beta} + g_{\alpha\beta}(\ \overset{(2k)}{S}{}^{\lambda}_{\mu}|^{\mu}_{\lambda} - \overset{(2k)}{S}{}^{\lambda}_{\lambda}|^{\mu}_{\mu}),\ k \geq 1 \qquad \textbf{(9R.8)}
$$

whose validity can be established by direct verification. If we introduce (9R.8) into (9.15) and rearrange, we obtain

$$
(1+\nu)[\ \overset{(2)}{S}{}_{\alpha\beta} + \Lambda h^2(-\frac{2}{21}\ \overset{(2)}{S}{}_{\alpha\beta} + \frac{1}{63}\ \overset{(4)}{S}{}_{\alpha\beta})|^{\lambda}_{\lambda}] - g_{\alpha\beta}(\nu\ \overset{(2)}{S}{}^{\lambda}_{\lambda} + \nu_*\ \overset{(2)}{\sigma}\ ) + \frac{1}{45}\Lambda_* h^4\ \overset{(2)}{S}{}^{\lambda}_{\mu}|^{\mu}_{\lambda\alpha\beta}
$$

$$+ h^2 \left[-\frac{2}{21} \left([\Lambda(1+\nu) - \nu_*] \overset{(2)}{S}{}^{\lambda}_{\lambda|\alpha\beta} + \Lambda(1+\nu)g_{\alpha\beta}(\overset{(2)}{S}{}^{\lambda}_{\mu}|^{\mu}_{\lambda} - \overset{(2)}{S}{}^{\lambda}_{\lambda}|^{\mu}_{\mu}) + \Lambda_* \overset{(2)}{\sigma}|_{\alpha\beta} \right) \right.$$

$$\left. + \frac{1}{63} \left([\Lambda(1+\nu) - \nu_*] \overset{(4)}{S}{}^{\lambda}_{\lambda|\alpha\beta} + \Lambda(1+\nu)g_{\alpha\beta}(\overset{(4)}{S}{}^{\lambda}_{\mu}|^{\mu}_{\lambda} - \overset{(4)}{S}{}^{\lambda}_{\lambda}|^{\mu}_{\mu}) + \Lambda_* \overset{(4)}{\sigma}|_{\alpha\beta} \right) \right]$$

$$= \frac{1}{3}\nu_* h^2 \overset{(0)}{S}{}^{\lambda}_{\lambda|\alpha\beta} - \frac{1}{6}\Lambda_* h^2 p^*|_{\alpha\beta} \qquad \textbf{(9R.9a)}$$

$$(1+\nu) \left[\overset{(2k)}{S}_{\alpha\beta} + \Lambda h^2 \left(\frac{\overset{(2k-2)}{S}_{\alpha\beta}}{(4k-3)(4k-1)} - \frac{2\overset{(2k)}{S}_{\alpha\beta}}{(4k-1)(4k+3)} + \frac{\overset{(2k+2)}{S}_{\alpha\beta}}{(4k+3)(4k+5)} \right)|^{\lambda}_{\lambda} \right]$$

$$- g_{\alpha\beta}[\nu \overset{(2k)}{S}{}^{\lambda}_{\lambda} + \nu_* \overset{(2k)}{\sigma}]$$

$$+ h^2 \left[\frac{[\Lambda(1+\nu) - \nu_*] \overset{(2k-2)}{S}{}^{\lambda}_{\lambda|\alpha\beta} + \Lambda(1+\nu)g_{\alpha\beta}(\overset{(2k-2)}{S}{}^{\lambda}_{\mu}|^{\mu}_{\lambda} - \overset{(2k-2)}{S}{}^{\lambda}_{\lambda}|^{\mu}_{\mu}) + \Lambda_* \overset{(2k-2)}{\sigma}|_{\alpha\beta}}{(4k-3)(4k-1)} \right.$$

$$- 2\frac{[\Lambda(1+\nu) - \nu_*] \overset{(2k)}{S}{}^{\lambda}_{\lambda|\alpha\beta} + \Lambda(1+\nu)g_{\alpha\beta}(\overset{(2k)}{S}{}^{\lambda}_{\mu}|^{\mu}_{\lambda} - \overset{(2k)}{S}{}^{\lambda}_{\lambda}|^{\mu}_{\mu}) + \Lambda_* \overset{(2k)}{\sigma}|_{\alpha\beta}}{(4k-1)(4k+3)}$$

$$\left. + \frac{[\Lambda(1+\nu) - \nu_*] \overset{(2k+2)}{S}{}^{\lambda}_{\lambda|\alpha\beta} + \Lambda(1+\nu)g_{\alpha\beta}(\overset{(2k+2)}{S}{}^{\lambda}_{\mu}|^{\mu}_{\lambda} - \overset{(2k+2)}{S}{}^{\lambda}_{\lambda}|^{\mu}_{\mu}) + \Lambda_* \overset{(2k+2)}{\sigma}|_{\alpha\beta}}{(4k+3)(4k+5)} \right]$$

$$= 0, \; k \geq 2 \qquad \textbf{(9R.9b)}$$

Equations (9R.3), (9R.5), (9R.7) and (9R.9) constitute the system of equations for the determination of the residual effects: they are to be solved subject to the edge conditions (9.22), with even index $n = 2k$.

In particular we note that the system of equations takes a considerably simplified form in the case of transverse inextensibility when $\Lambda_* = \nu_* = 0$.

10. The Problem of Bending

In the problem of bending the surface forces are applied anti-symmetrically, so that

$$p^+ = -p^- = \frac{1}{2}p, \; p^* = 0 \qquad \textbf{(10.1)}$$

while the specification of the stress vectors on the edge surface is such that, in their dependence on the thickness coordinate, the laminar components are odd while the transverse shear components are even, namely

$$\overset{(2k)}{T^{\alpha}}(\eta) = 0, \quad \overset{(2k)}{D}(\eta) = 0, \; k \geq 0. \tag{10.2}$$

The Legendre representations for the stress components have the form

$$\sigma^{\alpha\beta} = \sum_{k=0}^{\infty} \overset{(2k+1)}{S^{\alpha\beta}}(x^{\lambda})\, P_{2k+1}(t) \tag{10.3a}$$

$$\tau^{\alpha} = \sum_{k=0}^{\infty} \overset{(2k)}{\tau^{\alpha}}(x^{\lambda})\, P_{2k}(t) \tag{10.3b}$$

$$\sigma = \sum_{k=0}^{\infty} \overset{(2k+1)}{\sigma}(x^{\lambda})\, P_{2k+1}(t) \tag{10.3c}$$

in which the zero-th coefficients satisfy the pair of equations

$$\overset{(0)}{\tau^{\alpha}} = \frac{1}{3} h \overset{(1)}{S^{\beta\alpha}}\big|_{\beta} \tag{10.4}$$

$$2h \overset{(0)}{\tau^{\alpha}}\big|_{\alpha} + p = 0 \tag{10.5}$$

reproduced respectively from (4.21) and (4.22). The introduction of $\overset{(0)}{\tau^{\alpha}}$ from (10.4) into (10.5) leads to the equation satisfied by the principal laminar bending coefficients $\overset{(1)}{S^{\alpha\beta}}$, namely

$$\frac{2}{3} h^2 \overset{(1)}{S^{\lambda}_{\mu}}\big|^{\mu}_{\lambda} + p = 0 \tag{10.6}$$

already derived with different dummy indices as equation (4.25).

Before formulating the system of equations for the higher laminar coefficients in the expansion (10.3a), we first express the remaining coefficients in terms of them. Since relation (4.27) follows the general pattern expressed by formulae (4.7), we have for the higher coefficients in the expansion (10.3b)

$$\overset{(2k)}{\tau^{\alpha}} = h \left[\frac{\overset{(2k+1)}{S^{\beta\alpha}}\big|_{\beta}}{4k+3} - \frac{\overset{(2k-1)}{S^{\beta\alpha}}\big|_{\beta}}{4k-1} \right], \; k \geq 1 \tag{10.7}$$

for which the case $k = 0$ is covered by (10.4). Making a familiar change in the indexing on the right hand sides of relations (4.29), (4.31) and the general form (4.11) with odd index n, we have the system of formulae

$$\overset{(1)}{\sigma} = \frac{1}{35} h^2 \overset{(3)}{S}{}^{\lambda\mu}{}_{|\mu\lambda} + \frac{3}{5} p \tag{10.8a}$$

$$\overset{(3)}{\sigma} = h^2 \left[-\frac{2}{45} \overset{(3)}{S}{}^{\lambda\mu}{}_{|\mu\lambda} + \frac{1}{99} \overset{(5)}{S}{}^{\lambda\mu}{}_{|\mu\lambda} \right] - \frac{1}{10} p \tag{10.8b}$$

$$\overset{(2k+1)}{\sigma} = h^2 \left[\frac{\overset{(2k-1)}{S}{}^{\lambda\mu}{}_{|\mu\lambda}}{(4k-1)(4k+1)} - \frac{2\ \overset{(2k+1)}{S}{}^{\lambda\mu}{}_{|\mu\lambda}}{(4k+1)(4k+5)} + \frac{\overset{(2k+3)}{S}{}^{\lambda\mu}{}_{|\mu\lambda}}{(4k+5)(4k+7)} \right], \quad k \geq 2 \tag{10.8c}$$

for the transverse normal stress coefficients in the expansion (10.3c).

For the displacement components, we have the Legendre representations

$$u_\alpha = \sum_{k=0}^{\infty} \overset{(2k+1)}{u}_\alpha (x^\lambda)\, P_{2k+1}(t) \tag{10.9a}$$

$$w = \sum_{k=0}^{\infty} \overset{(2k)}{w} (x^\lambda)\, P_{2k}(t) \tag{10.9b}$$

in which only $\overset{(0)}{w}$ remains to be determined. With the transverse normal stress coefficients given by (10.8), relations (6.15a,c) render the laminar displacement coefficients in terms of the laminar stress coefficients, according to the formulae

$$\begin{aligned} \overset{(1)}{u}_\alpha = -h\, \overset{(0)}{w}|_\alpha + \frac{h^2}{E} \Bigg[& \frac{1}{15} \left(12\Lambda(1+\nu)\, \overset{(1)}{S}{}^{\lambda}{}_{\alpha|\lambda} - \nu_*\, \overset{(1)}{S}{}^{\lambda}{}_{\lambda|\alpha} + \Lambda_*\, \overset{(1)}{\sigma}|_\alpha \right) \\ & - \frac{1}{35} \left(2\Lambda(1+\nu)\, \overset{(3)}{S}{}^{\lambda}{}_{\alpha|\lambda} - \nu_*\, \overset{(3)}{S}{}^{\lambda}{}_{\lambda|\alpha} + \Lambda_*\, \overset{(3)}{\sigma}|_\alpha \right) \Bigg] \end{aligned} \tag{10.10a}$$

$$\begin{aligned} \overset{(2k+1)}{u}_\alpha = -\frac{h^2}{E} \Bigg[& \frac{2\Lambda(1+\nu)\, \overset{(2k-1)}{S}{}^{\lambda}{}_{\alpha|\lambda} - \nu_*\, \overset{(2k-1)}{S}{}^{\lambda}{}_{\lambda|\alpha} + \Lambda_*\, \overset{(2k-1)}{\sigma}|_\alpha}{(4k-1)(4k+1)} \\ & - 2\frac{2\Lambda(1+\nu)\, \overset{(2k+1)}{S}{}^{\lambda}{}_{\alpha|\lambda} - \nu_*\, \overset{(2k+1)}{S}{}^{\lambda}{}_{\lambda|\alpha} + \Lambda_*\, \overset{(2k+1)}{\sigma}|_\alpha}{(4k+1)(4k+5)} \\ & + \frac{2\Lambda(1+\nu)\, \overset{(2k+3)}{S}{}^{\lambda}{}_{\alpha|\lambda} - \nu_*\, \overset{(2k+3)}{S}{}^{\lambda}{}_{\lambda|\alpha} + \Lambda_*\, \overset{(2k+3)}{\sigma}|_\alpha}{(4k+5)(4k+7)} \Bigg], \quad k \geq 1 \end{aligned} \tag{10.10b}$$

while relation (6.8) with even index n yields the higher transverse displacement coefficients in the expansion (10.9b) in the form

$$\overset{(2k)}{w} = -\frac{h}{E}\left[\frac{\Lambda_*\overset{(2k+1)}{\sigma} - \nu_*\overset{(2k+1)}{S}{}^{\lambda}_{\lambda}}{4k+3} - \frac{\Lambda_*\overset{(2k-1)}{\sigma} - \nu_*\overset{(2k-1)}{S}{}^{\lambda}_{\lambda}}{4k-1}\right], \quad k \geq 1 \tag{10.11}$$

The corresponding series for the associated rotation components are

$$\psi = \sum_{k=0}^{\infty} \overset{(2k+1)}{\psi}(x^{\lambda})\, P_{2k+1}(t) \tag{10.12a}$$

$$\omega_{\alpha} = \sum_{k=0}^{\infty} \overset{(2k)}{\omega}_{\alpha}(x^{\lambda})\, P_{2k}(t) \tag{10.12b}$$

in which, from (6.17) the laminar elements are given by

$$\overset{(1)}{\psi} = \frac{h^2}{E}\Lambda(1+\nu)\left[\frac{2}{5}\left(\overset{(1)}{S}{}^{\lambda}_{1|\lambda 2} - \overset{(1)}{S}{}^{\lambda}_{2|\lambda 1}\right) - \frac{1}{35}\left(\overset{(3)}{S}{}^{\lambda}_{1|\lambda 2} - \overset{(3)}{S}{}^{\lambda}_{2|\lambda 1}\right)\right] \tag{10.13a}$$

$$\overset{(2k+1)}{\psi} = -\frac{h^2}{E}\Lambda(1+\nu)\left[\frac{\overset{(2k-1)}{S}{}^{\lambda}_{1|\lambda 2} - \overset{(2k-1)}{S}{}^{\lambda}_{2|\lambda 1}}{(4k-1)(4k+1)} - 2\,\frac{\overset{(2k+1)}{S}{}^{\lambda}_{1|\lambda 2} - \overset{(2k+1)}{S}{}^{\lambda}_{2|\lambda 1}}{(4k+1)(4k+5)} + \frac{\overset{(2k+3)}{S}{}^{\lambda}_{1|\lambda 2} - \overset{(2k+3)}{S}{}^{\lambda}_{2|\lambda 1}}{(4k+5)(4k+7)}\right], \quad k \geq 1 \tag{10.13b}$$

while for the transverse coefficients in the expansion (10.12b), we have, from (6.12),

$$\overset{(0)}{\omega}_{\alpha} = \frac{h}{E}\Lambda(1+\nu)\,\frac{1}{3}\,h\,\overset{(1)}{S}{}^{\lambda}_{\alpha|\lambda} - \overset{(0)}{w}\big|_{\alpha} \tag{10.14a}$$

$$\overset{(2k)}{\omega}_{\alpha} = \frac{h}{E}\left[\frac{\Lambda(1+\nu)\overset{(2k+1)}{S}{}^{\lambda}_{\alpha|\lambda} - \nu_*\overset{(2k+1)}{S}{}^{\lambda}_{\lambda|\alpha} + \Lambda_*\overset{(2k+1)}{\sigma}\big|_{\alpha}}{4k+3} - \frac{\Lambda(1+\nu)\overset{(2k-1)}{S}{}^{\lambda}_{\alpha|\lambda} - \nu_*\overset{(2k-1)}{S}{}^{\lambda}_{\lambda|\alpha} + \Lambda_*\overset{(2k-1)}{\sigma}\big|_{\alpha}}{4k-1}\right], \quad k \geq 1 \tag{10.14b}$$

The introduction of expressions (10.8a,b) for $\overset{(1)}{\sigma}$ and $\overset{(3)}{\sigma}$ into (7.4) leads to the following form of the equations for the determination of the mean displacement

$$-Eh\,\overset{(0)}{w}|_{\alpha\beta} = (1+\nu)\overset{(1)}{S}_{\alpha\beta} - g_{\alpha\beta}[\nu \overset{(1)}{S}{}^{\lambda}_{\lambda} + \nu_*(\tfrac{3}{5}p + \tfrac{1}{35}h^2 \overset{(3)}{S}{}^{\lambda|\mu}_{\mu|\lambda})]$$

$$-\frac{1}{15}h^2[6\Lambda(1+\nu)(\overset{(1)}{S}{}^{\lambda}_{\alpha|\lambda\beta} + \overset{(1)}{S}{}^{\lambda}_{\beta|\lambda\alpha}) - \nu_* \overset{(1)}{S}{}^{\lambda}_{\lambda|\alpha\beta} + \Lambda_*(\tfrac{3}{5}p + \tfrac{1}{35}h^2 \overset{(3)}{S}{}^{\lambda|\mu}_{\mu|\lambda})_{|\alpha\beta}]$$

$$+\frac{1}{35}h^2[\Lambda(1+\nu)(\overset{(3)}{S}{}^{\lambda}_{\alpha|\lambda\beta} + \overset{(3)}{S}{}^{\lambda}_{\beta|\lambda\alpha}) - \nu_* \overset{(3)}{S}{}^{\lambda}_{\lambda|\alpha\beta}$$

$$-\Lambda_*(\frac{1}{10}p + \frac{2}{45}h^2 \overset{(3)}{S}{}^{\lambda|\mu}_{\mu|\lambda} - \frac{1}{99}h^2 \overset{(5)}{S}{}^{\lambda|\mu}_{\mu|\lambda})_{|\alpha\beta}] \qquad \mathbf{(10.15)}$$

The equations for the residual laminar stress coefficients follow from the system (7.5) with odd index: writing the case $n = 3$ separately, we substitute for $\overset{(1)}{\sigma}$ from (10.18a) so that the system of equations takes the form

$$(1+\nu)\overset{(3)}{S}_{\alpha\beta} - g_{\alpha\beta}(\nu \overset{(3)}{S}{}^{\lambda}_{\lambda} + \nu_* \overset{(3)}{\sigma}) + \frac{1}{525}\Lambda_* h^4 \overset{(3)}{S}{}^{\lambda|\mu}_{\mu|\lambda\alpha\beta}$$

$$+h^2\Bigg[-\frac{2}{45}\Big(\Lambda(1+\nu)(\overset{(3)}{S}{}^{\lambda}_{\alpha|\lambda\beta} + \overset{(3)}{S}{}^{\lambda}_{\beta|\lambda\alpha}) - \nu_* \overset{(3)}{S}{}^{\lambda}_{\lambda|\alpha\beta} + \Lambda_* \overset{(3)}{\sigma}|_{\alpha\beta}\Big)$$

$$+\frac{1}{99}\Big(\Lambda(1+\nu)(\overset{(5)}{S}{}^{\lambda}_{\alpha|\lambda\beta} + \overset{(5)}{S}{}^{\lambda}_{\beta|\lambda\alpha}) - \nu_* \overset{(5)}{S}{}^{\lambda}_{\lambda|\alpha\beta} + \Lambda_* \overset{(5)}{\sigma}|_{\alpha\beta}\Big)\Bigg]$$

$$= -\frac{1}{15}\Lambda(1+\nu)h^2(\overset{(1)}{S}{}^{\lambda}_{\alpha|\lambda\beta} + \overset{(1)}{S}{}^{\lambda}_{\alpha|\lambda\beta}) + \frac{1}{15}\nu_* h^2 \overset{(1)}{S}{}^{\lambda}_{\lambda|\alpha\beta} - \frac{1}{25}\Lambda_* h^2 p|_{\alpha\beta} \qquad \mathbf{(10.16a)}$$

$$(1+\nu)\overset{(2k+1)}{S}_{\alpha\beta} - g_{\alpha\beta}(\nu \overset{(2k+1)}{S}{}^{\lambda}_{\lambda} + \nu_* \overset{(2k+1)}{\sigma})$$

$$+h^2\Bigg[\frac{\Lambda(1+\nu)(\overset{(2k-1)}{S}{}^{\lambda}_{\alpha|\lambda\beta} + \overset{(2k-1)}{S}{}^{\lambda}_{\beta|\lambda\alpha}) - \nu_* \overset{(2k-1)}{S}{}^{\lambda}_{\lambda|\alpha\beta} + \nu_* \overset{(2k-1)}{\sigma}|_{\alpha\beta}}{(4k-1)(4k+1)}$$

$$-2\frac{\Lambda(1+\nu)(\overset{(2k+1)}{S}{}^{\lambda}_{\alpha|\lambda\beta} + \overset{(2k+1)}{S}{}^{\lambda}_{\beta|\lambda\alpha}) - \nu_* \overset{(2k+1)}{S}{}^{\lambda}_{\lambda|\alpha\beta} + \Lambda_* \overset{(2k+1)}{\sigma}|_{\alpha\beta}}{(4k+1)(4k+5)}$$

$$+\frac{\Lambda(1+\nu)(\overset{(2k+3)}{S}{}^{\lambda}_{\alpha|\lambda\beta}+\overset{(2k+3)}{S}{}^{\lambda}_{\beta|\lambda\alpha})-\nu_*\overset{(2k+3)}{S}{}^{\lambda}_{\lambda|\alpha\beta}+\Lambda_*\overset{(2k+3)}{\sigma}|_{\alpha\beta}}{(4k+5)(4k+7)}\Bigg]=0,\ k\geq 1 \quad \textbf{(10.16b)}$$

We define the residual effects as those induced exclusively by the residual laminar stress coefficients $\{\overset{(2k+1)}{S}{}^{\alpha\beta}, k\geq 1\}$. The resolution of the field quantities into their principal and residual parts then follows from an inspection of the component representations in the infinite dimensional linear vector space induced by the expansion in terms of Legendre functions. For the transverse normal stress, we have

$$\sigma=\{\overset{(2k+1)}{\sigma}, 0\leq k\leq\infty\} \quad \textbf{(10.17)}$$

in which each component is the sum of a principal effect $\overset{(2k+1)}{\sigma}{}^{P}$ and a residual effect $\overset{(2k+1)}{\sigma}{}^{R}$ giving

$$\overset{(2k+1)}{\sigma}=\overset{(2k+1)}{\sigma}{}^{P}+\overset{(2k+1)}{\sigma}{}^{R} \quad \textbf{(10.18)}$$

From (10.18) we see that

$$\overset{(1)}{\sigma}{}^{P}=\frac{3}{5}p,\qquad \overset{(1)}{\sigma}{}^{R}=\frac{1}{35}h^2\overset{(3)}{S}{}^{\lambda}_{\mu}{}^{\lambda}_{|\lambda} \quad \textbf{(10.19a)}$$

$$\overset{(3)}{\sigma}{}^{P}=-\frac{1}{10}p,\quad \overset{(3)}{\sigma}{}^{R}=h^2[-\frac{2}{45}\overset{(3)}{S}{}^{\lambda\mu}_{\mu|\lambda}+\frac{1}{99}\overset{(5)}{S}{}^{\lambda\mu}_{\mu|\lambda}] \quad \textbf{(10.19b)}$$

$$\overset{(2k+1)}{\sigma}{}^{P}=0,\qquad \overset{(2k+1)}{\sigma}{}^{R}=\overset{(2k+1)}{\sigma},\ k\geq 2 \quad \textbf{(10.19c)}$$

so that, if we write

$$\sigma^{P}=\{\overset{(2k+1)}{\sigma}{}^{P}, 0\leq k\leq\infty\},\ \sigma^{R}=\{\overset{(2k+1)}{\sigma}{}^{R}, 0\leq k\leq\infty\} \quad \textbf{(10.20)}$$

then the standard convention for vector addition gives

$$\sigma=\sigma^{P}+\sigma^{R} \quad \textbf{(10.21)}$$

The corresponding resolution of the other field quantities follows the same pattern.

Letting $\overset{(R)}{W}$ denote the residual component of the mean transverse displacement, it follows from (10.15) that $\overset{(R)}{W}$ must satisfy the differential equations

$$-Eh\overset{(R)}{W}|_{\alpha\beta} = -\frac{1}{35}\nu_* h^2[g_{\alpha\beta}\overset{(3)}{S}{}^{\lambda}_{\mu}|^{\mu}_{\lambda} + \overset{(3)}{S}{}^{\lambda}_{\lambda|\alpha\beta}]$$

$$+\frac{1}{35}h^2[\Lambda(1+\nu)(\overset{(3)}{S}{}^{\lambda}_{\alpha|\lambda\beta} + \overset{(3)}{S}{}^{\lambda}_{\beta|\lambda\alpha}) - \frac{1}{9}\Lambda_* h^2(\overset{(3)}{S}{}^{\lambda}_{\mu}|^{\mu}_{\lambda} - \frac{1}{11}\overset{(3)}{S}{}^{\lambda}_{\mu}|^{\mu}_{\lambda})|_{\alpha\beta}] \qquad \textbf{(10.22)}$$

where we have combined terms in the second bracket. The system consisting of (10.16) together with (10.22), when complemented by the edge conditions (5.14c) with odd index n, namely

$$\overset{(2k+1)}{S}{}^{\beta\alpha}\mathfrak{n}_\alpha = \overset{(2k+1)}{T}{}^{\alpha}, \quad h\overset{(2k+1)}{S}{}^{\alpha\beta}|_\beta\mathfrak{n}_\alpha = \overset{(2k+1)}{D}, \; k \geq 1 \qquad \textbf{(10.23)}$$

constitute the boundary value problem for the determination of the residual effects: it is presumed that the principal problem has been already solved so that the terms involving the principal coefficients $\overset{(1)}{S}_{\alpha\beta}$ in (10.16a) have been calculated and thus appear as inhomogeneous terms.

We write W for the principal component of the mean transverse displacement so that

$$W = \overset{(0)}{W} + \overset{(R)}{W} \qquad \textbf{(10.24)}$$

and, by subtracting (10.22) from (10.15), we have

$$-EhW|_{\alpha\beta} = (1+\nu)\overset{(1)}{S}_{\alpha\beta} - g_{\alpha\beta}(\nu\overset{(1)}{S}{}^{\lambda}_{\lambda} + \frac{3}{5}\nu_* p)$$

$$-\frac{2}{5}\Lambda(1+\nu)h^2(\overset{(1)}{S}{}^{\lambda}_{\alpha|\lambda\beta} + \overset{(1)}{S}{}^{\lambda}_{\beta|\lambda\alpha}) + \frac{1}{15}\nu_* h^2\overset{(1)}{S}{}^{\lambda}_{\lambda|\alpha\beta} - \frac{3}{70}\Lambda_* h^2 p|_{\alpha\beta} \qquad \textbf{(10.25)}$$

which, when combined with the equilibrium equation (10.6) and the edge conditions (5.14b), namely

$$\overset{(1)}{S}{}^{\beta\alpha}\mathfrak{n}_\beta = \overset{(1)}{T}{}^{\alpha}, \quad h\overset{(1)}{S}{}^{\beta\alpha}|_\beta\mathfrak{n}_\alpha = \overset{(1)}{D} \qquad \textbf{(10.26)}$$

constitute the boundary value problem for the principal effects.

10P. The Principal Bending Problem

For the further analysis of the principal bending problem, governed by equations (10.25) and (10.6) we start by taking the trace of the tensor equation (10.25): if we substitute from (10.6) into the resulting relation, we find

$$-EhW|^{\alpha}_{\alpha} = (1-\nu)\overset{(1)}{S}{}^{\alpha}_{\alpha} + \frac{1}{15}\nu_* h^2 \overset{(1)}{S}{}^{\lambda}_{\lambda}|^{\alpha}_{\alpha} + [\Lambda(1+\nu)-\nu_*]\frac{6}{5}p - \frac{3}{70}\Lambda_* h^2 p|^{\alpha}_{\alpha} \qquad \textbf{(10P.1)}$$

A second relation is derived by considering the contravariant form of (10.25) and taking the repeated covariant derivative with respect to the variables x^{α} and x^{β}: after summing over the repeated indices, we multiply by h^2 and again utilize equation (10.6) to obtain

$$-Eh^3 W|^{\alpha\beta}_{\alpha\beta} = -\nu h^2 \overset{(1)}{S}{}^{\lambda}_{\lambda}|^{\alpha}_{\alpha} + \frac{1}{15}\nu_* h^4 \overset{(1)}{S}{}^{\lambda}_{\lambda}|^{\alpha\beta}_{\alpha\beta} - \frac{3}{2}(1+\nu)p$$

$$+ [\Lambda(1+\nu) - \frac{1}{2}\nu_*]\frac{6}{5}h^2 p|^{\alpha}_{\alpha} - \frac{3}{70}\Lambda_* h^4 p|^{\alpha\beta}_{\alpha\beta} \qquad \textbf{(10P.2)}$$

The consistency of relations (10P.1) and (10P.2) implies a compatibility relation for the principal stress components. If the Laplacian operator is applied to relations (10P.1), then on multiplication by h^2, there follows an alternative form of relation (10P.2) namely

$$-Eh^3 W|^{\alpha\beta}_{\alpha\beta} = (1-\nu)h^2 \overset{(1)}{S}{}^{\alpha}_{\alpha}|^{\beta}_{\beta} + \frac{1}{15}\nu_* h^4 \overset{(1)}{S}{}^{\lambda}_{\lambda}|^{\alpha\beta}_{\alpha\beta}$$

$$+ [\Lambda(1+\nu) - \nu_*]\frac{6}{5}h^2 p|^{\alpha}_{\alpha} - \frac{3}{70}\Lambda_* h^4 p|^{\alpha\beta}_{\alpha\beta} \qquad \textbf{(10P.3)}$$

The comparison of (10P.2) with (10P.3) shows that the principal stresses must satisfy the compatibility equation

$$h^2 \overset{(1)}{S}{}^{\lambda}_{\lambda}|^{\alpha}_{\alpha} = -\frac{3}{2}(1+\nu)p + \frac{3}{5}\nu_* h^2 p|^{\alpha}_{\alpha} \qquad \textbf{(10P.4)}$$

From the introduction of this latter relation into (10P.1), we have the formula

$$(1-\nu)\overset{(1)}{S}{}^{\alpha}_{\alpha} = -EhW|^{\alpha}_{\alpha} - [(\Lambda - \frac{1}{12}\nu_*)(1+\nu) - \nu_*]\frac{6}{5}p + (\Lambda_* - \frac{14}{15}\nu_*^2)\frac{3}{70}h^2 p|^{\alpha}_{\alpha} \qquad \textbf{(10P.5)}$$

expressing the trace of the laminar stress tensor in terms of the principal component of the mean displacement and the surface pressures. If we now apply the Laplacian operator to (10P.5) and then substitute from (10P.4) into the left hand side of the resulting equation, we obtain the differential equation to be satisfied by W, namely

$$Eh^3 W|^{\alpha\beta}_{\alpha\beta} = \frac{3}{2}(1-\nu^2)p - (\Lambda - \frac{7}{12}\nu_*)(1+\nu)\frac{6}{5}h^2 p|^{\alpha}_{\alpha} + (\Lambda_* - \frac{14}{15}\nu_*^2)\frac{3}{70}h^4 p|^{\alpha\beta}_{\alpha\beta} \qquad \textbf{(10P.6)}$$

Next by substituting for $\overset{(1)}{S}{}^{\lambda}_{\lambda}$ from (10P.5) into the tensor equation (10.25), we see that the latter may be written

$$(1-\nu^2)[\overset{(1)}{S}_{\alpha\beta} - \frac{2}{5}\Lambda h^2(\overset{(1)}{S}{}^{\lambda}_{\alpha|\lambda\beta} + \overset{(1)}{S}{}^{\lambda}_{\beta|\lambda\alpha})]$$

$$= -Eh[(1-\nu)W|_{\alpha\beta} + \nu g_{\alpha\beta}W|^{\lambda}_{\lambda} - \frac{1}{15}\nu_* h^2 W|^{\lambda}_{\lambda\alpha\beta}]$$

$$-[\nu(\Lambda - \frac{1}{12}\nu_*) - \frac{1}{2}\nu_*](1+\nu)g_{\alpha\beta}\frac{6}{5}p$$

$$+[\nu_*(\Lambda - \frac{1}{12}\nu_*)(1+\nu) - \nu_*^2 + \frac{15}{28}\Lambda_*(1-\nu)]\frac{2}{25}h^2 p|_{\alpha\beta}$$

$$+\frac{3}{70}(\Lambda_* - \frac{14}{15}\nu_*^2)[\nu g_{\alpha\beta}h^2 p|^{\lambda}_{\lambda} - \frac{1}{15}\nu_* h^4 p|^{\lambda}_{\lambda\alpha\beta}] \quad \textbf{(10P.7)}$$

Moreover from the identity

$$\overset{(1)}{S}{}^{\lambda}_{\alpha|\lambda\beta} + \overset{(1)}{S}{}^{\lambda}_{\beta|\lambda\alpha} = \overset{(1)}{S}_{\alpha\beta}|^{\lambda}_{\lambda} + \overset{(1)}{S}{}^{\lambda}_{\lambda|\alpha\beta} + g_{\alpha\beta}(\overset{(1)}{S}{}^{\lambda\mu}_{\mu}|_{\lambda} - \overset{(1)}{S}{}^{\lambda\mu}_{\lambda}|_{\mu}) \quad \textbf{(10P.8)}$$

which holds for any symmetric second order tensor in the plane, it follows from the introduction of $\overset{(1)}{S}{}^{\lambda}_{\lambda}$ from (10P.5), of $\overset{(1)}{S}{}^{\lambda\mu}_{\mu}|_{\lambda}$ from (10.6) and of $\overset{(1)}{S}{}^{\lambda\mu}_{\lambda}|_{\mu}$ from (10P.4) that

$$h^2(\overset{(1)}{S}{}^{\lambda}_{\alpha|\lambda\beta} + \overset{(1)}{S}{}^{\lambda}_{\beta|\lambda\alpha}) = h^2\overset{(1)}{S}_{\alpha\beta}|^{\lambda}_{\lambda} - \frac{Eh^3}{1-\nu}W|^{\lambda}_{\lambda\alpha\beta} + g_{\alpha\beta}\frac{3}{2}[(2+\nu)p - \frac{2}{5}\nu_* h^2 p|^{\lambda}_{\lambda}]$$

$$-\frac{1}{1-\nu}[\left((\Lambda - \frac{1}{12}\nu_*)(1+\nu) - \nu_*\right)\frac{6}{5}h^2 p_{|\alpha\beta} - (\Lambda_* - \frac{14}{15}\nu_*^2)\frac{3}{70}h^4 p|^{\lambda}_{\lambda\alpha\beta}]$$

(10P.9)

If we now substitute from (10P.9) into (10P.7), then the latter system of relations takes the form

$$\overset{(1)}{S}_{\alpha\beta} - \frac{2}{5}h^2\overset{(1)}{S}_{\alpha\beta}|^{\lambda}_{\lambda}$$

$$= -\frac{Eh}{1-\nu^2}[(1-\nu)W_{|\alpha\beta} + \nu g_{\alpha\beta}W|^{\lambda}_{\lambda} + \frac{2}{5}\left(\Lambda(1+\nu) - \frac{1}{6}\nu_*\right)h^2 W|^{\lambda}_{\lambda\alpha\beta}]$$

$$+ [\Lambda\left(2 - \frac{\nu(1+\nu)}{1-\nu}\right) + \frac{\nu_*}{1-\nu}(1 + \frac{1}{6}\nu)]g_{\alpha\beta}\frac{3}{5}p$$

$$-[\left(6\frac{\Lambda}{1-\nu} - \frac{\nu}{1-\nu^2}\right)\left((1+\nu)(\Lambda - \frac{1}{12}\nu_*) - \nu_*\right) - \frac{15}{28}\frac{\Lambda_*}{1+\nu}]\frac{2}{25}h^2 p_{|\alpha\beta}$$

$$-[\Lambda\nu_* - \frac{5}{28}\frac{\nu}{1-\nu^2}(\Lambda_* - \frac{14}{15}\nu_*^2)]g_{\alpha\beta}\frac{6}{25}h^2 p|^\lambda_\lambda$$

$$+(\Lambda_* - \frac{14}{15}\nu_*^2)(\frac{\Lambda}{1-\nu} - \frac{1}{6}\frac{\nu_*}{1-\nu^2})\frac{3}{175}h^4 p|^\lambda_{\lambda\alpha\beta} \qquad \textbf{(10P.10)}$$

Equations (10P.6) and (10P.10), with the supplementary equations (10P.4), (10P.5) and (10.6) describe the principal bending problem of plate theory: they are to be solved subject to the edge conditions (10.26).

To facilitate comparison as well as for greater ease in the subsequent discussion, we now consider the simplified form taken by these equations when the surface pressures vanish. In that case, equation (10P.6) reduces to

$$W|^{\alpha\beta}_{\alpha\beta} = 0 \qquad \textbf{(10P.11)}$$

and, in the notation of (4.32b), relations (10P.10) become

$$M_{\alpha\beta} - \frac{2}{5}\Lambda h^2 M_{\alpha\beta}|^\lambda_\lambda = -\frac{2Eh^3}{3(1-\nu^2)}[(1-\nu)W_{|\alpha\beta} + \nu g_{\alpha\beta}W|^\lambda_\lambda$$

$$+\frac{2}{5}\left(\Lambda(1+\nu) - \frac{1}{6}\nu_*\right)h^2 W|^\lambda_{\lambda\alpha\beta}] \qquad \textbf{(10P.12)}$$

while the supplementary equations (10P.4), (10P.5) and (10.6) read

$$M^\lambda_\lambda|^\alpha_\alpha = 0 \qquad \textbf{(10P.13)}$$

$$M^\alpha_\alpha = -\frac{2Eh^3}{3(1-\nu)}W|^\alpha_\alpha \qquad \textbf{(10P.14)}$$

$$M^\lambda_\mu|^\mu_\lambda = 0 \qquad \textbf{(10P.15)}$$

respectively. Except for the effects due to transverse normal stress, as reflected in the terms with ν_* on the right of (10P.12), the above equations are identical in form with those of Reissner's theory of Plate Bending.

If we consider the asymptotic form of (10P.12) for small h, by substituting the form of the right hand side for $M_{\alpha\beta}$ in the differentiated term on the left, then on noting (10P.11) and rearranging, we find

$$M_{\alpha\beta} \simeq -\frac{2Eh^3}{3(1-\nu^2)}[(1-\nu)W|_{\alpha\beta} + \nu g_{\alpha\beta}W|^{\lambda}_{\lambda} + (\tfrac{4}{5}\Lambda - \tfrac{1}{15}\nu_*)h^2 W|^{\lambda}_{\lambda\alpha\beta}] \qquad \textbf{(10P.16)}$$

In the above system, if we replace (10P.12) by (10P.16), we have the equivalent of the "Thick" Plate theory of Michel-Love. There is a discrepancy in the effect due to transverse normal stress: what appears as $-\frac{1}{15}\nu_*$ in (10P.16) corresponds to $+\frac{1}{10}\nu_*$ in the Michel-Love derivation. However if the latter is reformulated in terms of the mean displacement, rather than for the midsurface displacement of the original, this discrepancy disappears.

Finally we note that if we omit the terms with h^2 in the bracket on the right of (10P.16), we have

$$M_{\alpha\beta} \simeq -\frac{2Eh^2}{3(1-\nu^2)}[(1-\nu)W|_{\alpha\beta} + \nu g_{\alpha\beta}W|^{\lambda}_{\lambda}] \qquad \textbf{(10P.17)}$$

which, if used instead of (10P.12) yields the classical plate theory of Kirchhoff.

Both the Michel-Love theory and the Kirchhoff theory are contracted "interior" theories to be solved subject to the contracted Kirchhoff edge conditions: only by retaining the form of the left hand side of (10P.12) intact can the three independent conditions be satisfied.

10R: The Residual Bending Problem

Taking the trace of the tensor equation (10.22), we have

$$-Eh\overset{(R)}{W}|^{\alpha}_{\alpha} = -\frac{1}{35}\nu_* h^2[2\overset{(3)}{S}{}^{\lambda}_{\mu}|^{\mu}_{\lambda} + \overset{(3)}{S}{}^{\lambda}_{\lambda}|^{\alpha}_{\alpha}]$$
$$+\frac{1}{35}h^2[2\Lambda(1+\nu)\overset{(3)}{S}{}^{\lambda}_{\alpha}|^{\alpha}_{\lambda} - \frac{1}{9}\Lambda_* h^2(\overset{(3)}{S}{}^{\lambda}_{\mu}|^{\mu}_{\lambda} - \frac{1}{11}\overset{(5)}{S}{}^{\lambda}_{\mu}|^{\mu}_{\lambda})|^{\alpha}_{\alpha}] \qquad \textbf{(10R.1)}$$

If we also consider the contravariant form of (10.22) and, after taking the repeated covariant derivative, sum over the repeated indices, we obtain

$$-Eh\overset{(R)}{W}|^{\alpha\beta}_{\alpha\beta} = -\frac{1}{35}\nu_* h^2[\overset{(3)}{S}{}^{\lambda}_{\mu}|^{\mu\alpha}_{\lambda\alpha} + \overset{(3)}{S}{}^{\lambda}_{\lambda}|^{\alpha\beta}_{\alpha\beta}]$$
$$+\frac{1}{35}h^2[2\Lambda(1+\nu)\overset{(3)}{S}{}^{\lambda}_{\alpha}|^{\alpha\beta}_{\lambda\beta} - \frac{1}{9}\Lambda_* h^2(\overset{(3)}{S}{}^{\lambda}_{\mu}|^{\mu}_{\lambda} - \frac{1}{11}\overset{(5)}{S}{}^{\lambda}_{\mu}|^{\mu}_{\lambda})|^{\alpha\beta}_{\alpha\beta}] \qquad \textbf{(10R.2)}$$

For the consistency of (10R.1) and (10R.2), we have

$$\overset{(3)}{S}{}^{\lambda}{}_{\mu}\big|^{\mu\alpha}_{\lambda\alpha} = 0 \qquad \textbf{(10R.3)}$$

and hence also

$$Eh^3 \overset{(R)}{W}\big|^{\alpha\beta}_{\alpha\beta} = \frac{1}{35} h^4 [\nu_* \overset{(3)}{S}{}^{\lambda}_{\lambda} - \frac{1}{99} \Lambda_* \overset{(5)}{S}{}^{\lambda}{}_{\mu}\big|^{\mu}_{\lambda}]\big|^{\alpha\beta}_{\alpha\beta} \qquad \textbf{(10R.4)}$$

so that, except for an additive biharmonic function, we have

$$E\overset{(R)}{W} = \frac{1}{35} h[\nu_* \overset{(3)}{S}{}^{\lambda}_{\lambda} - \frac{1}{99} \Lambda_* h^2 \overset{(5)}{S}{}^{\lambda}{}_{\mu}\big|^{\mu}_{\lambda}] \qquad \textbf{(10R.5)}$$

The omitted biharmonic term has already been accounted a principal effect.

We next write the relations resulting from taking the trace of the tensor equations (10.16), namely

$$(1-\nu)\overset{(3)}{S}{}^{\alpha}_{\alpha} - 2\nu_* \overset{(3)}{\sigma}$$

$$+ h^2[-\frac{2}{45}\left(2\Lambda(1+\nu)\overset{(3)}{S}{}^{\lambda}{}_{\alpha}\big|^{\alpha}_{\lambda} - \nu_* \overset{(3)}{S}{}^{\lambda}{}_{\lambda}\big|^{\alpha}_{\alpha} + \Lambda_* \overset{(3)}{\sigma}\big|^{\alpha}_{\alpha}\right)$$

$$+ \frac{1}{99}\left(2\Lambda(1+\nu)\overset{(5)}{S}{}^{\lambda}{}_{\alpha}\big|^{\alpha}_{\lambda} - \nu_* \overset{(5)}{S}{}^{\lambda}{}_{\lambda}\big|^{\alpha}_{\alpha} + \Lambda_* \overset{(5)}{\sigma}\big|^{\alpha}_{\alpha}\right)]$$

$$= \frac{1}{10}(1+\nu)(2\Lambda - \nu_*)p + \frac{1}{25}(\nu_*^2 - \Lambda_*)h^2 p\big|^{\alpha}_{\alpha} \qquad \textbf{(10R.6a)}$$

$$(1-\nu)\overset{(2k+1)}{S}{}^{\alpha}_{\alpha} - 2\nu_* \overset{(2k+1)}{\sigma}$$

$$+ h^2\left[\frac{2\Lambda(1+\nu)\overset{(2k-1)}{S}{}^{\lambda}{}_{\alpha}\big|^{\alpha}_{\lambda} - \nu_* \overset{(2k-1)}{S}{}^{\lambda}{}_{\lambda}\big|^{\alpha}_{\alpha} + \Lambda_* \overset{(2k-1)}{\sigma}\big|^{\alpha}_{\alpha}}{(4k-1)(4k+1)}\right.$$

$$-2\,\frac{2\Lambda(1+\nu)\overset{(2k+1)}{S}{}^{\lambda}{}_{\alpha}\big|^{\alpha}_{\lambda} - \nu_* \overset{(2k+1)}{S}{}^{\lambda}{}_{\lambda}\big|^{\alpha}_{\alpha} + \Lambda_* \overset{(2k+1)}{\sigma}\big|^{\alpha}_{\alpha}}{(4k+1)(4k+5)}$$

$$\left. + \frac{2\Lambda(1+\nu)\overset{(2k+3)}{S}{}^{\lambda}{}_{\alpha}\big|^{\alpha}_{\lambda} - \nu_* \overset{(2k+3)}{S}{}^{\lambda}{}_{\lambda}\big|^{\alpha}_{\alpha} + \Lambda_* \overset{(2k+3)}{\sigma}\big|^{\alpha}_{\alpha}}{(4k+5)(4k+7)}\right] = 0,\ k \geq 2 \qquad \textbf{(10R.6b)}$$

where in (10R.6a) we have noted (10R.3) and also substituted from (10.6) and (10P.4) on the right.

We obtain an alternative system of these relations if we take the repeated covariant derivative of the contravariant form (10.16) and sum over the repeated indices: we find

$$(1+\nu)\overset{(3)}{S}{}^{\alpha}{}_{\beta}|^{\beta}_{\alpha} - \nu \overset{(3)}{S}{}^{\lambda}{}_{\lambda}|^{\alpha}_{\alpha} - \nu_* \overset{(3)}{\sigma}|^{\alpha}_{\alpha}$$

$$+ h^2\Big[-\frac{2}{45}\big(-\nu_* \overset{(3)}{S}{}^{\lambda}{}_{\lambda}|^{\alpha\beta}_{\alpha\beta} + \Lambda_* \overset{(3)}{\sigma}|^{\alpha\beta}_{\alpha\beta}\big) + \frac{1}{99}\big(2\Lambda \overset{(5)}{S}{}^{\lambda}{}_{\alpha}|^{\alpha\beta}_{\lambda\beta} - \nu_* \overset{(5)}{S}{}^{\lambda}{}_{\lambda}|^{\alpha}_{\alpha} + \Lambda_* \overset{(5)}{\sigma}|^{\alpha\beta}_{\alpha\beta}\big)\Big]$$

$$= \frac{1}{10}(1+\nu)(2\Lambda-\nu_*)p|^{\alpha}_{\alpha} + \frac{1}{25}(\nu_*^2 - \Lambda_*)h^2 p|^{\alpha\beta}_{\alpha\beta} \qquad \textbf{(10R.7a)}$$

$$(1+\nu)\overset{(2k+1)}{S}{}^{\alpha}{}_{\beta}|^{\beta}_{\alpha} - \nu \overset{(2k+1)}{S}{}^{\lambda}{}_{\lambda}|^{\alpha}_{\alpha} - \nu_* \overset{(2k+1)}{\sigma}|^{\alpha}_{\alpha}$$

$$+h^2\left[\frac{2\Lambda(1+\nu)\overset{(2k-1)}{S}{}^{\lambda}{}_{\alpha}|^{\alpha\beta}_{\lambda\beta} - \nu_* \overset{(2k-1)}{S}{}^{\lambda}{}_{\lambda}|^{\alpha\beta}_{\alpha\beta} + \Lambda_* \overset{(2k-1)}{\sigma}|^{\alpha\beta}_{\alpha\beta}}{(4k-1)(4k+1)}\right.$$

$$-2\,\frac{2\Lambda(1+\nu)\overset{(2k+1)}{S}{}^{\lambda}{}_{\alpha}|^{\alpha\beta}_{\lambda\beta} - \nu_* \overset{(2k+1)}{S}{}^{\lambda}{}_{\lambda}|^{\alpha\beta}_{\alpha\beta} + \Lambda_* \overset{(2k+1)}{\sigma}|^{\alpha\beta}_{\alpha\beta}}{(4k+1)(4k+5)}$$

$$\left.+\frac{2\Lambda(1+\nu)\overset{(2k+3)}{S}{}^{\lambda}{}_{\alpha}|^{\alpha\beta}_{\lambda\beta} - \nu_* \overset{(2k+3)}{S}{}^{\lambda}{}_{\lambda}|^{\alpha\beta}_{\alpha\beta} + \Lambda_* \overset{(2k+3)}{\sigma}|^{\alpha\beta}_{\alpha\beta}}{(4k+5)(4k+7)}\right] = 0,\ k \geq 2 \qquad \textbf{(10R.7b)}$$

where in (10R.7a) we have again substituted from (10.6), (10P.4) and (10R.3). The consistency of each equation of the system (10R.6) with the corresponding equation in (10R.7) then leads to the sequence of compatibility relations

$$(1+\nu)\overset{(2k+1)}{S}{}^{\alpha}{}_{\beta}|^{\beta}_{\alpha} - \overset{(2k+1)}{S}{}^{\lambda}{}_{\lambda}|^{\alpha}_{\alpha} + \nu_* \overset{(2k+1)}{\sigma}|^{\alpha}_{\alpha} = 0 \qquad \textbf{(10R.8)}$$

With the $\overset{(2k+1)}{\sigma}$, $k \geq 1$, given by (10.8b,c) and also noting (10R.3), the two systems of equations (10R.6) and (10R.8) are for the simultaneous determination of the two sequences of invariants $\{\overset{(2k+1)}{S}{}^{\lambda}_{\lambda},\ k \geq 1\}$, and $\{\overset{(2k+1)}{S}{}^{\lambda}{}_{\mu}|^{\mu}_{\lambda},\ k \geq 1\}$.

If we utilize the identities

$$\overset{(2k+1)}{S}{}^{\lambda}{}_{\alpha|\lambda\beta} + \overset{(2k+1)}{S}{}^{\lambda}{}_{\beta|\lambda\alpha} = \overset{(2k+1)}{S}{}_{\alpha\beta}|^{\lambda}_{\lambda} + \overset{(2k+1)}{S}{}^{\lambda}{}_{\lambda|\alpha\beta} + g_{\alpha\beta}\big(\overset{(2k+1)}{S}{}^{\lambda}{}_{\mu}|^{\mu}_{\lambda} - \overset{(2k+1)}{S}{}^{\lambda}{}_{\lambda}|^{\mu}_{\mu}\big),\ k \geq 0 \qquad \textbf{(10R.9)}$$

in relations (10.16), we obtain the explicit form of the equations for the determination of the individual components, namely

$$
(1+\nu)\Big[\overset{(3)}{S}_{\alpha\beta}+\Lambda h^{2}\Big(-\frac{2}{45}\overset{(3)}{S}_{\alpha\beta}+\frac{1}{99}\overset{(5)}{S}_{\alpha\beta}\Big)\Big|^{\lambda}_{\lambda}\Big]-g_{\alpha\beta}\Big(\nu\overset{(3)}{S}{}^{\lambda}_{\lambda}+\nu_{*}\overset{(3)}{\sigma}\Big)+\frac{1}{525}\Lambda_{*}h^{4}\overset{(3)}{S}{}^{\lambda\mu}_{\mu|\lambda\alpha\beta}
$$

$$
+h^{2}\Bigg[-\frac{2}{45}\Big([\Lambda(1+\nu)-\nu_{*}]\overset{(3)}{S}{}^{\lambda}_{\lambda|\alpha\beta}+\Lambda(1+\nu)g_{\alpha\beta}\big(\overset{(3)}{S}{}^{\lambda\mu}_{\mu}\big|^{\lambda}_{\lambda}-\overset{(3)}{S}{}^{\lambda}_{\lambda}\big|^{\mu}_{\mu}\big)+\Lambda_{*}\overset{(3)}{\sigma}\big|_{\alpha\beta}\Big)
$$

$$
+\frac{1}{99}\Big([\Lambda(1+\nu)-\nu_{*}]\overset{(5)}{S}{}^{\lambda}_{\lambda|\alpha\beta}+\Lambda(1+\nu)g_{\alpha\beta}\big(\overset{(5)}{S}{}^{\lambda\mu}_{\mu}\big|^{\lambda}_{\lambda}-\overset{(5)}{S}{}^{\lambda}_{\lambda}\big|^{\mu}_{\mu}\big)+\Lambda_{*}\overset{(5)}{\sigma}\big|_{\alpha\beta}\Big)\Bigg]
$$

$$
=-\frac{1}{15}\Lambda(1+\nu)h^{2}\overset{(1)}{S}_{\alpha\beta}\big|^{\lambda}_{\lambda}-\frac{1}{15}[\Lambda(1+\nu)-\nu_{*}]h^{2}\overset{(1)}{S}{}^{\lambda}_{\lambda|\alpha\beta}
$$

$$
-\frac{1}{10}\Lambda(1+\nu)g_{\alpha\beta}\Big(\nu p-\frac{2}{5}\nu_{*}h^{2}p\big|^{\lambda}_{\lambda}\Big)-\frac{1}{25}\Lambda_{*}h^{2}p\big|_{\alpha\beta}\qquad\textbf{(10R.10a)}
$$

$$
(1+\nu)\Bigg[\overset{(2k+1)}{S}_{\alpha\beta}+\Lambda h^{2}\Bigg(\frac{\overset{(2k-1)}{S}_{\alpha\beta}}{(4k-1)(4k+1)}-\frac{2\overset{(2k+1)}{S}_{\alpha\beta}}{(4k+1)(4k+5)}+\frac{\overset{(2k+3)}{S}_{\alpha\beta}}{(4k+5)(4k+7)}\Bigg)\Bigg|^{\lambda}_{\lambda}\Bigg]
$$

$$
-g_{\alpha\beta}\Big[\nu\overset{(2k+1)}{S}{}^{\lambda}_{\lambda}+\nu_{*}\overset{(2k+1)}{\sigma}\Big]
$$

$$
+h^{2}\Bigg[\frac{[\Lambda(1+\nu)-\nu_{*}]\overset{(2k-1)}{S}{}^{\lambda}_{\lambda|\alpha\beta}+\Lambda(1+\nu)g_{\alpha\beta}\big(\overset{(2k-1)}{S}{}^{\lambda\mu}_{\mu}\big|^{\lambda}_{\lambda}-\overset{(2k-1)}{S}{}^{\lambda}_{\lambda}\big|^{\mu}_{\mu}\big)+\Lambda_{*}\overset{(2k-1)}{\sigma}\big|_{\alpha\beta}}{(4k-1)(4k+1)}
$$

$$
-2\frac{[\Lambda(1+\nu)-\nu_{*}]\overset{(2k+1)}{S}{}^{\lambda}_{\lambda|\alpha\beta}+\Lambda(1+\nu)g_{\alpha\beta}\big(\overset{(2k+1)}{S}{}^{\lambda\mu}_{\mu}\big|^{\lambda}_{\lambda}-\overset{(2k+1)}{S}{}^{\lambda}_{\lambda}\big|^{\mu}_{\mu}\big)+\Lambda_{*}\overset{(2k+1)}{\sigma}\big|_{\alpha\beta}}{(4k+1)(4k+5)}
$$

$$
+\frac{[\Lambda(1+\nu)-\nu_{*}]\overset{(2k+3)}{S}{}^{\lambda}_{\lambda|\alpha\beta}+\Lambda(1+\nu)g_{\alpha\beta}\big(\overset{(2k+3)}{S}{}^{\lambda\mu}_{\mu}\big|^{\lambda}_{\lambda}-\overset{(2k+3)}{S}{}^{\lambda}_{\lambda}\big|^{\mu}_{\mu}\big)+\Lambda_{*}\overset{(2k+3)}{\sigma}\big|_{\alpha\beta}}{(4k+5)(4k+7)}\Bigg]
$$

$$
=0,\ k\geq 2\qquad\textbf{(10R.10b)}
$$

where on the right of (10R.10a), we have substituted from (10.6) and (10P.4).

Equations (10R.3), (10R.6), (10R.8) and (10R.10) constitute the system of equations for the determination of the residual effects: they are to be solved subject to the edge conditions (10.23), with odd index $n = 2k+1$. Again, we note that the system assumes a simplified form in the case of transverse inextensibility, when $\Lambda_* = \nu_* = 0$.

Background Survey

The history of plate theory, which is almost coincident with that of the General Theory of Elasticity, was launched from the particular interest in plate vibration in the early nineteenth century. This beginning gave to the bending problem a preferred position which it has since consistently retained. The greater emphasis on the bending case is understandable for the principal problems, since the problem of stretching does not exhibit a comparable variety of interesting mathematical features. However, the extension of this bias in the residual problems can only be explained by association since, in the analysis of the residual effects, both problems are equally challenging.

The approximate description of stress and strain in a stretched plate was discussed by some of the earlier authors, notably by Clebsch [10] in 1862, whose analysis extended the earlier work of Kirchhoff and Gehring, for which reference may be found in the treatise by Kirchhoff [21(b)]. While this theory is substantially equivalent to the principal stretching problem, its formulation as the generalized theory of plane stress appears to date from the later work of Filon [11] in 1903. The formal similarity of the latter theory to the two-dimensional theory of plane stress of Airy [2], suggested the stress-function representation leading to the reduction to the biharmonic equation. Apart from the rather restricted refinements suggested by Reissner [33(a)], and later by Reiss and Locke [32], the residual problems seems to have received but scant attention.

The history of the theory of plate bending begins with the publication of the work of Germain [15] in 1821, who deduced the biharmonic equation for the transverse deflection from a two-dimensional extension of the moment-curvature hypothesis of Euler. This work preceded the formulation of the General Equations of Elasticity by Cauchy [8] later in the same decade. On the basis of this general theory, distinct derivations, based on power series expansions, of the contracted two-dimensional equations, were subsequently proposed by Cauchy [8(c)] and by Poisson [30], from which it was soon recognized that the order of the system of equations was not adequate for the satisfaction of the three independent edge conditions, presumed necessary to a two-dimensional formulation. These difficulties remained unresolved until the appearance of the classic paper of Kirchhoff [21(a)] in 1850.

As a two-dimensional adaptation of Coulomb's kinematic assumption for beams, Kirchhoff adopted the hypothesis, that normals to the undeformed midplane deform without extension into normals to the deformed midplane, whose consequences he then combined with the Principle of Virtual Work. The variational procedure then yielded the two-dimensional equations in the form of the Eulerian minimizing conditions, while the appropriately contracted edge conditions emerged as the natural boundary conditions. The order of the differential equations was now correct for the satisfaction of the two Kirchhoff conditions. Comparing the latter with the three originally proposed, one remains unchanged while the remaining two are merged in the second Kirchhoff condition. It is not surprising that this feature of Kirchhoff's derivation aroused considerable interest. The theory was further developed in the subsequent works of both Kirchhoff [21(b)] and Clebsch [10].

It appears that most of the lingering reluctance to accept the Kirchhoff contraction was removed by the work of Kelvin and Tait [20], where it was shown how the merging of two physical conditions in one Kirchhoff condition could be interpreted in terms of the "equivalence of statically equipollent systems" as prescribed by Saint Venant's Principle. Specific difficulties of the (Kirchhoff) theory were subsequently considered by a number of investigators, notably Boussinesq [7] and Lamb [24]. The larger problems associated with relating the classical two-dimensional theory to the three-dimensional theory were discussed by Hadamard [19]. Reference to prior investigations of these aspects of the problem may be found in this paper of Hadamard, who seems to have been alone in recognizing the value of the interesting work of Levy [25].

Following the semi-inverse procedure of Saint Venant, Michel [28] in 1900 proposed a refinement of the classical theory, which, in the somewhat simpler form in which it was reproduced by Love [26], has been called the "Theory of Moderately Thick Plates." A rather specialized refinement for a specific thick plate problem was derived by Mesnager [27]. A more general procedure, based on series expansions in terms of a thickness parameter, was developed by G.D. Birkhoff [4] in connection with obtaining refinements in the theory of plates of varying thickness. This latter method was further extended in the specific cases considered by C.A. Garabedian [14]. Both the Michel-Love and Birkhoff theories, although they offer corrections to the classical formulae, nevertheless reflect their asymptomatic nature by their retention of the Kirchhoff contraction in the edge conditions.

In 1944 there appeared the significant papers of Reissner [33(b), (c)] which finally revealed the boundary layer nature of the effects suppressed by the Kirchhoff contraction. By incorporating the transverse shear stress into the expression for the strain energy, Reissner derived a set of equations permitting the satisfaction of three independent edge conditions. Similar results were later obtained independently by Bolle [6] whose method was based on the relaxation of the Kirchhoff hypothesis so as to admit transverse shear strain. Except for the numerically unimportant terms due to transverse normal stress, Reissner's equations are formally identical with equations (10P.12) to (10P.15). We have seen how the Kirchhoff and Michel-Love theories then appear respectively as one- and two-term representations, asymptomatically valid in the interior. Other aspects of the theory were discussed in the later paper of Reissner [33d] and in the alternative derivation by Green [17(a)], which may also be found in the book by Green and Zerna [18]. A neat summary of Reissner's theory together with a brief consideration of the questions it left unanswered may be found in the discussion by Goldenweizer [16(a)].

Meantime an alternative approach to the edge-effect phenomena revealed by Reissner's theory was proposed by Friedrichs [12]. This method, based on the matching of inner and outer asymptotic expansions, was furher developed in the later treatment of Friedrichs and Dressler [13] and also in the papers of Goldenweizer [16(b)] where references to related work by other investigators in the Soviet Union may be found. Here also reference should be made to the work of Nordgren [29] and Simmonds [34] wherein the method of the hypercircle is applied to derive estimates on the error in the theory of plate bending in terms of the thickness-wavelenth ratio.

The idea of Legendre expansions, originally developed by Cicala [9] has been adapted by Poniatovski [31], and more recently by Krenk [22], to the variational method followed by Reissner. However, it appears that, in the former application, the retention of more terms does not lead to any modification in the final form of Reissner's equations, whereas in the latter adaptation, the incorporation of the effect of transverse extensibility necessitates a re-interpretation of the displacement function appearing in Reissner's theory.

In previous considerations of the residual bending problem, the residual effects have been treated as higher order corrections to the principal effects within the three-dimensional theory. Such a process of successive refinement has been dealt with in some detail in the analysis of Green [17(b)], based on Fourier expansions and also in the later work of Alblas [3]. We have already noted the simplication effected in the equations governing the residual effects by the assumption of transverse inextensibility. Again in the context of determining refinements to the principal effects, this simpler case has been studied in the papers of Kromm [23] and also in the subsequent analysis by Boal and Reissner [5], who also discuss the contraction of the edge conditions required by the assumption of transverse inextensibility.

The residual problems as formulated in Sections 9R and 10R involve infinite systems of partial differential equations. Certain analytic questions posed by such systems have been treated in the studies of Agmon, Douglis and Nirenberg [1], where references to other related material may be found.

References

1. Agmon, S., Douglis, A., and Nirenberg, L.,
(a) Comm. Pure Appl. Math., vol. 12, 1959.
(b) Comm. Pure Appl. Mech., vol. 14, 1964.
2. Airy, G. B., Phil. Trans. Roy. Soc., vol. 153, 1863.
3. Alblas, J. B., *Theorie von de dreidimensionale spanningstoestand in een doorboorde plaat,* Amsterdam, Holland: 1957. H. J. Paris, 1957.
4. Birkhoff, G. D., Phil. Mag., (ser. 6), vol. 43, 1922.
5. Boal, J. L. and Reissner, E., J. Math and Phys., vol. 39, 1960.
6. Bolle, L., Bull. Tech. Suisse Roman, vol. 73, 1947.
7. Boussinesq, J., (a) J. de Math. (Liouville) (ser. 2) t. 16, 1871.
(b) J. de Math. (Liouville) (ser. 3) t. 5, 1879.
8. Cauchy, A. L., (a) Bull. des Sciences de la Société philomathique, 1823.
(b) Exercises de Mathématique, vol. 2, 1827.
(c) Exercises de Mathématique, vol. 3, 1828.
9. Cicala, P., Giornio Genio Civile, vol. 97, 1959.
10. Clebsch, A., *Theorie der Elasticitat fester Korper,* Leipsig, 1862.
11. Filon, L. N. G., Phil. Trans. Roy. Soc. (ser. A), vol. 201, 1903.
12. Friedrichs, K. O.,
(a) *Reissner Anniversary Volumen,* Edwards, Ann Arbor, 1949.
(b) Proc. Symp. in Appl. Math., vol. 3, McGraw-Hill, New York, 1950.
13. Friedrichs, K. O. and Dressler, R. F., Comm. Pure Appl. Math., vol. 14, 1961.
14. Garabedian, C. A., (a) Amer. Math. Soc. Trans., vol. 25, 1923.
(b) Paris C. R., t. 178, 1924.
15. Germain, S., *Recherches sur la théorie des surfaces élastiques,* Paris, 1821.
16. Goldenweizer, A. L., (a) On Reissner's Theory of Bending of Plates, (Eng. tr. of Russian original), NASA Rep. No. TTF-27, 1960.
(b) PMM (Appl Math. and Mech.), vol. 26, 1962.
17. Green, A. E., (a) Quart. Appl. Math., vol. 7, 1949.
(b) Phil. Trans. Roy. Soc., ser. A, vol. 240, 1949.
18. Green, A. E. and Zerna, W., *Theoretical Elasticity,* Oxford, 1954.
19. Hadamard, J., Trans. Amer. Math. Soc., vol. 3, 1902.
20. Kelvin, (Lord) and Tait, P. G., *Natural Philosophy,* 1st Ed., Oxford, 1867. 2nd Ed., Cambridge, 1883.
21. Kirchhoff, G., (a) J. f. Math. (Crelle), Bd. 40, 1850.
(b) Nostesungen uber Math. Physik, Mechanik (3rd Ed. Leipzig 1883).
22. Krenk, S., J. Appl. Mech., vol. 48, 1981.
23. Kromm, A., (a) Ingenieur Archiv., vol. 21, 1953.
(b) Zeit. angew. Math. Mech., vol. 35, 1955.
24. Lamb, H., Proc. London Math Soc., vol. 21, 1890.
25. Levy, M., J. Math Pures Appl., t. 3, 1877.
26. Love, A. E. H., *A Treatise on the Mathematical Theory of Elasticity,* Cambridge, 4th Ed., 1927.

27. Mesnager, A., Paris, C. R., t. 164, 1917.
28. Michel, J. H., Proc. London Math. Soc., vol. 31, 1900.
29. Nordgren, R. P., (a) Quart. Appl. Math., vol. 28, pp 587-595, 1971
(b) Quart. Appl. Math., vol. 29, pp 551-556, 1972
30. Poisson, S. D., Paris, Mem. de l'Acad., t. 8, 1829.
31. Poniatovskii, V. V., PMM (Appl. Math. and Mech.), vol. 26, 1962.
32. Reiss, E. L. and Locke, S., Quart. Appl. Math., vol. 19, 1961.
33. Reissner, E., (a) Proc. 15th Bull. Eastern Photoelasticity Conference, Boston, 1942.
(b) J. Math. and Phys., vol. 23, 1944.
(c) J. Appl. Mech., vol. 12, 1944.
(d) Quart. Appl. Math., vol. 5, 1947.
34. Simmonds, J. G., Quart. Appl. Math., vol. 29, pp 439-447, 1971.

"Shell Theory is hard."
Anonymous

Chapter Three

Shell Theory – A First Approximation

Introduction

The configuration of a plate, namely, a three-dimensional figure cut from a right cylinder by two mutually reflecting surfaces symmetrically placed with respect to a plane normal to the generators of the cylinder, is generalized in the shell by replacing the plane of symmetry by an arbitrary base surface, which, by analogy, is termed the midsurface. For a figure defined on the base surface by one or more simple closed curves, which we collectively call the edge-curve, the ruled surface, generated by the normals to the midsurface along the edge-curve, defines a region of space. Introducing the two faces, namely two surfaces mutually reflecting with respect to the midsurface, so that, on every normal to the latter, the intercept between the faces is bisected by the midsurface, then the portion of the ruled surface lying between the faces will be referred to as the edge-surface. The figure enclosed by the edge-surface and the two faces is a shell and for any point on the midsurface the normal intercept between the faces measures the shell-thickness at that point.

In the case of shells whose midsurface curvature variation is uniformly moderate, if also the maximum thickness is sufficiently small compared with the minimum radius of midsurface curvature so that each point of the shell is safely free from proximity to possible singular points, we say that the shell is thin. Moreover, we shall see that the square of this ratio of thickness to radius of curvature furnishes the small parameter, in terms of which, we effect an approximate integration, with respect to the thickness coordinate, of the governing equations. For thin shells, a first approximation is adequate.

The shell will be related to the general coordinate system natural to a configuration generated by the normal translation of the segment of a surface. The reference system will then consist of the system of laminar surfaces parallel to the midsurface together with the family of transversals normal thereto. The elements of the field quantities lying in the laminar surfaces will be called laminar components while those associated with the normal to the midsurface will be termed transverse effects.

The mechanical behavior of the shell material is described by the system of Cauchy-Green constitutive relations and we consider the shell so constructed that the mechanical characteristics associated with the transverse direction are distinct from those associated with directions in the laminar surfaces. We shall confine our analysis to the stress boundary value problem and, in order to focus on the factors of primary interest, we shall restrict our attention to the homogeneous shell of constant thickness, free from the action of any body force field, subjected to a purely normal pressure on the faces, but on whose edge-surface we consider the applied stress distribution in its most general form.

Resolving the edge stress distribution into its principal part consisting of the thickness-integrated stress-resultants and stress-couples, and its subsidiary part characterized by the fact that as a distribution it is self-equilibrating along any generator, then the principal problem of shell theory will be formulated in terms of the effects induced by the former while the residual problem will be concerned with the influence of the latter. The resolution is most conveniently effected through the Legendre representations for the stress quantities, which also facilitates the (approximate) integration of the differential equations with respect to the thickness coordinate. In the resulting integrated form we have an approximate two-dimensional formulation of the governing equations in which the thickness-averaged displacements are the only displacement quantities that appear explicitly. By making an appropriate decomposition of these mean displacements into their principal and residual parts, we effect the detachment of the system of equations describing the residual effects from the equations satisfied by the principal effects. A corresponding separation in the boundary conditions yields the uncoupled formulation of the principal problem together with a detached formulation of the residual problem.

Beyond its formulation we shall not concern ourselves with any further discussion of the residual problem. The subsequent analysis of the principal problem requires considerable manipulation extending the procedure previously followed in the reduction of the principal problems of plate theory. Finally, we derive the contracted form assumed by the equations governing the principal problem, when we introduce the simplifying approximation associated with the so-called "interior" description, valid, except in the edge-zone where the edge-effect due to transverse shear is significant. This corresponds to the Kirchoff asymptotic form of the equations of plate bending and has associated with it an analogous Kirchoff-type contraction of the boundary conditions.

The objectives of the analyses may now be stated under three main headings:

1) the clarification of the distinction between the principal and residual effects;
2) the separate formulation of the principal and residual boundary value problem;
3) the derivation of the constitutive relations for shells, adequate for describing the principal ("interior") effects in the region beyond the edge-zone.

In Section One we describe the coordinate system and derive the relevant tensor formulae: besides expressing the Christoffel symbols for the three-dimensional system in terms of the surface elements, we also relate the description of the edge-surface to the corresponding surface specification of the edge-curve. This is followed by an appendage giving the form taken by the expressions for the Christoffel symbols when the approximation for thin shells is introduced. The three-dimensional boundary value problem is formulated in Section Two, to which we also add the approximate form of the constitutive relations adequate for thin shells. With the introduction of the normalizing transformation in Section Three, we apply it to the exact form of the equilibrium equations and boundary conditions and to the approximate form of the stress-displacement relations for a reformulation of the problem.

From the integration of the equilibrium equations with respect to the thickness coordinate, we derive in Section Four the Legendre series representations for the stress components: the application of the face conditions then yields the equations satisfied by the leading coefficients which include the classical equilibrium equations of shell theory. By the introduction of the thin shell approximation into the relations connecting the coefficients in these series, we are led to the approximate expressions for the elements in the representations for the transverse stresses, in terms of the coefficients appearing in the series for the laminar stresses. Relating these representations to the expansions for the edge distributions applied at the edge-surface we effect, in Section Five, a reformulation of these conditions in terms of the laminar stress coefficients.

From the transverse integration of the approximate stress-displacement relations, we obtain, in Section Six, the associated Legendre representations for the displacement components, followed, in Section Seven, by the final two dimensional formulation of the equations involving only the laminar stress coefficients and the mean displacements. Having effected the detachment of the residual boundary value problem in Section Eight, we consider, in Section Nine, the complementary formulation of the boundary value problem for the principal effects and perform the reduction of the governing equations as suggested by the corresponding procedures followed in plate theory. In Section Ten, we introduce the Kirchhoff-type approximation to derive the set of constitutive relations adequate for a description of the principal effects in the interior region, valid at a sufficient distance from the edge.

We conclude with a brief survey of some of the previous work on shell theory together with a list of references.

1. The Coordinate System

The intrinsic geometry of a surface may be investigated in the reference frame of a general two dimensional coordinate system in terms of which the metric characterizing the surface is defined. Since, in the analysis of shell theory, it will be necessary to consider the midsurface within the framework of the imbedding three dimensional Euclidean space, we shall relate the metric elements of the former to the physical elements of the latter.

Referred to a general two dimensional Gaussian coordinate system $[x^\alpha, \alpha = 1,2]$

we may write for the position vector to an arbitrary point on the midsurface

$$\underset{\sim}{r} = \underset{\sim}{r}(x^{\alpha}) = \underset{\sim}{r}(x^{1},x^{2}) \tag{1.1}$$

By partial differentiation we introduce the base vectors

$$\underset{\sim}{g}_{\alpha} = \underset{\sim}{r}_{,\alpha} = \frac{\partial \underset{\sim}{r}}{\partial x^{\alpha}}, \quad \alpha = 1,2 \tag{1.2}$$

the scalar products of which yield the components of the symmetric covariant metric tensor

$$g_{\alpha\beta} = \underset{\sim}{g}_{\alpha} \cdot \underset{\sim}{g}_{\beta}, \quad \alpha,\beta = 1,2 \tag{1.3}$$

also called the first fundamental form of the surface with determinant

$$g = \det[g_{\alpha\beta}]_{1}^{2} = g_{11}g_{22} - g_{12}^{2}\,. \tag{1.4}$$

The related contravariant metric tensor is defined in the standard manner: using the Kronecker delta notation and with summation implied over repeated indices, the system of equations

$$g_{\alpha\lambda}g^{\lambda\beta} = \delta_{\alpha}^{\beta} \tag{1.5}$$

determine the components of the contravariant metric tensor $g^{\alpha\beta}$ in the form

$$g^{11} = \frac{g_{22}}{g}, \quad g^{22} = \frac{g_{11}}{g}, \quad g^{12} = -\frac{g_{12}}{g} \tag{1.6}$$

From these we define the system of conjugate base vectors by setting

$$\underset{\sim}{g}^{\alpha} = g^{\alpha\lambda}\underset{\sim}{g}_{\lambda} \tag{1.7}$$

from which it follows that

$$\underset{\sim}{g}_{\alpha} = g_{\alpha\lambda}\underset{\sim}{g}^{\lambda}, \quad \underset{\sim}{g}_{\alpha} \cdot \underset{\sim}{g}^{\beta} = \delta_{\alpha}^{\beta} \tag{1.8a,b}$$

With g given by (1.4), the associated permutation tensor $\epsilon_{\alpha\beta}$ has components

$$\epsilon_{11} = \epsilon_{22} = 0\,, \quad \epsilon_{12} = -\epsilon_{21} = \sqrt{g} \tag{1.9}$$

while the elements of the corresponding contravariant tensor $\epsilon^{\alpha\beta}$, defined by

$$\epsilon^{\alpha\beta} = g^{\alpha\lambda}g^{\beta\mu}\epsilon_{\lambda\mu} \tag{1.10}$$

are readily shown to be

$$\epsilon^{11} = \epsilon^{22} = 0 \,, \quad \epsilon^{12} = -\epsilon^{21} = \frac{1}{\sqrt{g}} \tag{1.11}$$

and therefore the identity

$$g^{\alpha\beta} = \epsilon^{\alpha\lambda}\epsilon^{\beta\mu} g_{\lambda\mu} \tag{1.12}$$

expresses relations (1.6) in tensor form.

The Christoffel symbols of the first and second kinds are defined as in the planar configuration, namely

$$[\alpha,\beta:\mu] = \frac{1}{2}[g_{\alpha\mu,\beta} + g_{\beta\mu,\alpha} - g_{\alpha\beta,\mu}] \tag{1.13a}$$

$$\left\{ {\lambda \atop \alpha\,\beta} \right\} = g^{\lambda\mu}[\alpha,\beta:\mu] = \frac{1}{2} g^{\lambda\mu}[g_{\alpha\mu,\beta} + g_{\beta\mu,\alpha} - g_{\alpha\beta,\mu}] \tag{1.13b}$$

from the trace of which there follows, by noting the form for the derivative of a determinant

$$\left\{ {\lambda \atop \alpha\,\lambda} \right\} = \frac{1}{2g}\frac{\partial g}{\partial x^{\alpha}} \tag{1.14}$$

giving

$$\frac{\partial}{\partial x^{\alpha}}\sqrt{g} = \sqrt{g}\left\{ {\lambda \atop \alpha\,\lambda} \right\} \tag{1.15}$$

Corresponding to the resolution formulae in the planar case, we shall see that, for the surface, the Christoffel symbols account for the projection, on the tangent plane, of the partial derivatives of the base vectors. In particular, by differentiation of relation (1.3) and forming appropriate combinations it can be readily verified that

$$\underset{\sim}{g}_{\alpha,\beta} \cdot \underset{\sim}{g}_{\mu} = [\alpha,\beta:\mu] \tag{1.16}$$

It then follows from relations (1.8) that

$$\underset{\sim}{g}_{\alpha,\beta} \cdot \underset{\sim}{g}^{\lambda} = \left\{ {\lambda \atop \alpha\,\beta} \right\} \,, \quad \underset{\sim}{g}^{\alpha}{}_{,\beta} \cdot \underset{\sim}{g}_{\lambda} = -\left\{ {\alpha \atop \beta\,\lambda} \right\} \tag{1.17a,b}$$

These latter formulae (1.16) and (1.17) reflect the fact that the intrinsic geometry is analyzed exclusively in terms of the first fundamental form.

For the fuller description it is necessary to introduce a second characteristic measure, that relates the surface to the imbedding Euclidean space. At each point of the surface the base vectors define a tangent plane, the normal to which is given by

$$\underset{\sim}{n}(x^{\alpha}) = \frac{\underset{\sim}{g}_1 \times \underset{\sim}{g}_2}{|\underset{\sim}{g}_1 \times \underset{\sim}{g}_2|} \tag{1.18}$$

The infinitesimal vector representing the instantaneous variation of the unit normal lies in the tangent plane: this is seen analytically by noting that from

$$\underset{\sim}{n} \cdot \underset{\sim}{n} = 1 \tag{1.19}$$

we derive

$$\underset{\sim}{n} \cdot \underset{\sim}{n}_{,\alpha} = 0 \tag{1.20}$$

Hence, since the system $[\underset{\sim}{g}_{\alpha}, \underset{\sim}{n}]$ constitutes a vector triad in the enclosing space, we may write for the resolution of the derivatives of the normal*

$$\underset{\sim}{n}_{,\alpha} = -b^{\lambda}_{\alpha}\, \underset{\sim}{g}_{\lambda} \tag{1.21}$$

where the b^{λ}_{α} are the components of a mixed tensor reflecting the curvature through its effect on the variation of the unit normal. Moreover, differentiation of the first of the orthogonality conditions

$$\underset{\sim}{g}_{\alpha} \cdot \underset{\sim}{n} = 0, \quad \underset{\sim}{g}^{\alpha} \cdot \underset{\sim}{n} = 0 \tag{1.22a,b}$$

yields the alternate form

$$\underset{\sim}{g}_{\alpha,\beta} \cdot \underset{\sim}{n} = -\underset{\sim}{g}_{\alpha}\, \underset{\sim}{n}_{,\beta} = g_{\alpha\lambda}\, b^{\lambda}_{\beta} = b_{\alpha\beta} \tag{1.23}$$

for the corresponding covariant components. Combining (1.17a) and (1.23) we obtain the resolution

$$\underset{\sim}{g}_{\alpha,\beta} = \left\{ {\lambda \atop \alpha\,\beta} \right\} \underset{\sim}{g}_{\lambda} + b_{\alpha\beta}\, \underset{\sim}{n} \tag{1.24}$$

for the derivatives of the base vectors. Similarly the result of differentiating the second orthogonality condition (1.22b) when combined with (1.17b) gives the corresponding resolution

$$\underset{\sim}{g}^{\alpha}{}_{,\beta} = -\left\{ {\alpha \atop \lambda\,\beta} \right\} \underset{\sim}{g}^{\lambda} + b^{\alpha}_{\beta}\, \underset{\sim}{n} \tag{1.25}$$

* These are the relations of Weingarten.

for the derivatives of the conjugate system.

The tensor $b_{\alpha\beta}$, whose components give the resolution, along the normal, of the derivatives of the base vectors, is called the second fundamental form and has associated with it the classical surface invariants, namely the mean and Gaussian curvatures. The mean curvature H is related to the trace by the formula

$$2H = g^{\lambda\mu}\, b_{\lambda\mu} = b_{\lambda}^{\lambda} = b_{1}^{1} + b_{2}^{2} \tag{1.26}$$

while the Gaussian curvature K is given in terms of the determinant by the relation

$$gK = \det[b_{\alpha\beta}]_{1}^{2} = b_{11}\, b_{22} - b_{12}\, b_{21} \tag{1.27}$$

or equivalently

$$K = b_{1}^{1}\, b_{2}^{2} - b_{2}^{1}\, b_{1}^{2} \tag{1.27*}$$

where we have noted formula (1.4) for g. Moreover, it can be readily verified that the relation

$$\epsilon^{\alpha\lambda}\, \epsilon^{\beta\mu}\, b_{\lambda\mu} + b^{\alpha\beta} = 2Hg^{\alpha\beta} \tag{1.28}$$

holds between the covariant and contravariant components.

The fact that the tensor $b_{\alpha\beta}$ constitutes the second fundamental form of a surface implies certain restrictions on the components. Recalling that

$$\underset{\sim}{g}_{\alpha,\beta} = \underset{\sim}{r}_{,\alpha\beta} = \underset{\sim}{r}_{,\beta\alpha} = \underset{\sim}{g}_{\beta,\alpha} \tag{1.29}$$

we see, from the defining relations (1.23), that the second fundamental form is a symmetric tensor, namely

$$b_{\alpha\beta} = b_{\beta\alpha} \tag{1.30}$$

Furthermore, from the observation that

$$\underset{\sim}{n}_{,\alpha\beta} = \underset{\sim}{n}_{,\beta\alpha} \tag{1.31}$$

there follows a second symmetry condition*

$$b_{\alpha|\beta}^{\lambda} = b_{\beta|\alpha}^{\lambda} \tag{1.32}$$

which with the first condition (1.30) implies for the covariant components

$$b_{\alpha\lambda|\beta} = b_{\beta\lambda|\alpha} = b_{\alpha\beta|\lambda} \tag{1.33}$$

Finally, we note that the celebrated theorem of Gauss, showing the invariant K to

* These relations are associated with the name of Codazzi and in the Soviet literature with that of Peterson.

be an intrinsic characteristic of the surface, expresses the determinant of the second fundamental form in terms of the elements of the first fundamental form.

For quantities defined in a general metric space the covariant derivative measures the intrinsic variation of the field elements and is identical in form with that for a Euclidean space. Accordingly, if we use a single stroke to denote covariant differentiation on the surface with base vectors $\underset{\sim}{g}_{\alpha}$ and metric $g_{\alpha\beta}$, then for the vector field

$$\underset{\sim}{a} = a^{\lambda} \underset{\sim}{g}_{\lambda} = a_{\lambda} \underset{\sim}{g}^{\lambda} \tag{1.34}$$

we have for the covariant derivatives of the respective components

$$a^{\lambda}|_{\alpha} = a^{\lambda}{}_{,\alpha} + \left\{ {\lambda \atop \mu\,\alpha} \right\} a^{\mu} \tag{1.35a}$$

$$a_{\lambda|\alpha} = a_{\lambda,\alpha} - \left\{ {\mu \atop \lambda\,\alpha} \right\} a_{\mu} \tag{1.35b}$$

while for the components of a second order tensors $a^{\alpha\beta}$ and $a_{\alpha\beta}$ the corresponding formulae read

$$a^{\alpha\beta}|_{\lambda} = a^{\alpha\beta}{}_{,\lambda} + \left\{ {\alpha \atop \mu\,\lambda} \right\} a^{\mu\beta} + \left\{ {\beta \atop \mu\,\lambda} \right\} a^{\alpha\mu} \tag{1.36a}$$

$$a_{\alpha\beta|\lambda} = a_{\alpha\beta,\lambda} - \left\{ {\mu \atop \beta\,\lambda} \right\} a_{\alpha\mu} - \left\{ {\mu \atop \alpha\,\lambda} \right\} a_{\mu\beta} \tag{1.36b}$$

However, in the expression for the partial derivative of a vector field there now appears an additional term due to the curvature. Taking the derivative of the vector (1.34), we insert the resolution formulae (1.24) and (1.25), and using the notation of (1.35), we obtain

$$\underset{\sim}{a}{}_{,\alpha} = a^{\lambda}|_{\alpha} \underset{\sim}{g}_{\lambda} + b_{\alpha\lambda} a^{\lambda} \underset{\sim}{n} \tag{1.37a}$$

$$= a_{\lambda|\alpha} \underset{\sim}{g}^{\lambda} + b^{\lambda}_{\alpha} a_{\lambda} \underset{\sim}{n} \tag{1.37b}$$

emphasizing how the normal component of the variation is a product of the second fundamental form.

We next outline the relevant formulae for a general coordinate system $[x^{i}, i = 1,2,3]$ in a three dimensional Euclidean space. Writing

$$\underset{\sim}{R} = \underset{\sim}{R}(x^{i}) = \underset{\sim}{R}(x^{1},x^{2},x^{3}) \tag{1.38}$$

for the position vector to an arbitrary point we derive the base vectors

$$\underset{\sim}{G}_i = \underset{\sim}{R}_{,i} = \frac{\partial \underset{\sim}{R}}{\partial x^i} \quad , \quad i = 1,2,3 \tag{1.39}$$

and form the covariant components of the metric tensor G_{ij} by setting

$$G_{ij} = \underset{\sim}{G}_i \cdot \underset{\sim}{G}_j \tag{1.40}$$

with determinant

$$G = \det[G_{ij}]_1^3 \tag{1.41}$$

The contravariant metric tensor is determined from the system of equation

$$G_{ik}\, G^{kj} = \delta_i^j \tag{1.42}$$

where the Kronecker symbol on the right denotes the unit matrix: then for the conjugate system of base vectors we have

$$\underset{\sim}{G}^i = G^{ik}\, \underset{\sim}{G}_k \tag{1.43}$$

and hence also the relations

$$\underset{\sim}{G}_i = G_{ik}\, \underset{\sim}{G}^k , \quad \underset{\sim}{G}_i \cdot G^j = \delta_i^j \tag{1.44}$$

In the three dimensional covariant permutation tensor ϵ_{ijk}, associated with the metric G_{ij}, the non-vanishing components are given by

$$\epsilon_{123} = \epsilon_{231} = \epsilon_{312} = \sqrt{G} \tag{1.45a}$$

$$\epsilon_{132} = \epsilon_{321} = \epsilon_{213} = -\sqrt{G} \tag{1.45b}$$

By noting the standard expansion of a determinant as the sum of products of its elements, it can be easily checked that in the corresponding contravariant tensor ϵ^{ijk} the non-vanishing elements are

$$\epsilon^{123} = \epsilon^{231} = \epsilon^{312} = \frac{1}{\sqrt{G}} \tag{1.46a}$$

$$\epsilon^{132} = \epsilon^{321} = \epsilon^{213} = -\frac{1}{\sqrt{G}} \tag{1.46b}$$

Then for a pair of vectors $\underset{\sim}{A}_{(1)}$ and $\underset{\sim}{A}_{(2)}$ with contravariant components $a^i_{(1)}$ and $a^i_{(2)}$ respectively, the associated cross product is defined as the vector A^* whose covariant components a_i^* are given by

$$a_i^* = \epsilon_{ijk}\, a_{(1)}^j\, a_{(2)}^k \tag{1.47}$$

For the Christoffel symbols of the first and second kinds, we write, respectively

$$\Gamma_{jk:i} = \frac{1}{2}\,[G_{ij,k} + G_{ik,j} - G_{jk,i}] \tag{1.48a}$$

$$\Gamma_{jk}^i = G^{il}\,\Gamma_{jk:l} = \frac{1}{2}\,G^{il}\,[G_{lj,k} + G_{lk,j} - G_{jk,l}] \tag{1.48b}$$

and the latter are the components of the derivatives of the base vectors, namely

$$\underset{\sim}{G}_{j,k} = \Gamma_{jk}^i\,\underset{\sim}{G}_i \tag{1.49}$$

In the case of an arbitrary vector field $\underset{\sim}{A}$ with resolution

$$\underset{\sim}{A} = a^i\,\underset{\sim}{G}_i = a_i\,\underset{\sim}{G}^i \tag{1.50}$$

partial differentiation with respect to one of coordinate variables yields

$$\underset{\sim}{A}_{,j} = a^i\|_j\,\underset{\sim}{G}_i = a_{i\|j}\,\underset{\sim}{G}^i \tag{1.51}$$

where we use a double stroke to distinguish covariant differentiation in the space with metric G_{ij}: namely, we have written

$$a^i\|_j = a^i_{,j} + \Gamma_{jk}^i\,a^k, \quad a_{i\|j} = a_{i,j} - \Gamma_{ij}^k\,a_k \tag{1.52a,b}$$

while for a second order tensor with components a^{ij}, contravariant in both indices, we would write

$$a^{ij}_{\ \ \|k} = a^{ij}_{\ \ ,k} + \Gamma_{lk}^i\,a^{lj} + \Gamma_{lk}^j\,a^{il} \tag{1.53}$$

as the defining formula for the covariant derivative

For the shell configuration, the natural frame of reference is the general conoidal coordinate system, consisting of the general Gaussian system $[x^\alpha, \alpha = 1,2]$ augmented by the transverse coordinate z normal to the midsurface. If we specialize the general system $[\mathrm{x}^i, i = 1,2,3]$, by identifying the first two coordinates with those of the Gaussian system $[x^\alpha, \alpha = 1,2]$ and equating the third with the transverse coordinate, namely

$$\mathrm{x}^\alpha = x^\alpha,\ \alpha = 1,2,\ \mathrm{x}^3 = z \tag{1.54}$$

then in place of (1.38) we write for the position vector to an arbitrary point

$$\underline{R} = \underline{R}(x^i) = \underline{R}(x^\alpha, z) \tag{1.55}$$

Observing that the latter is the resultant of adding the appropriate transverse segment to the position vector of the corresponding point of projection on the midsurface, namely

$$\underline{r}(x^\alpha) = \underline{R}_0 = \underline{R}(x^\alpha, 0) \tag{1.56}$$

it follows that the vector (1.54) has the resolution

$$\underline{R} = \underline{R}(x^i) = \underline{R}(x^\alpha, z) = \underline{r}(x^\alpha) + z\underline{n}(x^\alpha) \tag{1.57}$$

which enables us to relate the elements of the spatial coordinate system to the fundamental forms defined on the surface.

Recalling the defining relations (1.39), we differentiate (1.57), and noting formulae (1.2) and (1.21), we obtain for the base vectors of the conoidal coordinate system

$$\underline{G}_\alpha = \underline{g}_\alpha - zb^\lambda_\alpha \underline{g}_\lambda, \quad \alpha = 1,2: \quad \underline{G}_3 = \underline{n} \tag{1.58a:b}$$

and hence for the components of the metric tensor G_{ij} we have

$$G_{\alpha\beta} = g_{\alpha\beta} - 2zb_{\alpha\beta} + z^2 b^\lambda_\alpha b_{\lambda\beta}, \quad \alpha,\beta = 1,2 \tag{1.59a}$$

$$G_{13} = G_{23} = 0: \quad G_{33} = 1 \tag{1.59b:c}$$

The coefficient of z^2 in (1.59a) is sometimes called the third fundamental form: in fact it can be resolved as a combination of the first and second forms. We, therefore, obtain an alternate representation for the metric tensor in terms of the surface elements, namely

$$G_{\alpha\beta} = g_{\alpha\beta}(1 - z^2K) - 2zb_{\alpha\beta}(1 - zH), \quad \alpha,\beta = 1,2 \tag{1.60a}$$

$$G_{\alpha 3} = 0, \quad \alpha = 1,2: \quad G_{33} = 1 \tag{1.60b:c}$$

by introducing the tensor relation

$$b^\lambda_\alpha b_{\lambda\beta} = 2Hb_{\alpha\beta} - Kg_{\alpha\beta} \tag{1.61}$$

whose validity may be established by direct verification.

It follows from relations (1.60b:c) that the determinant of the metric tensor is given by

$$G = G_{11}G_{22} - G_{12}^2 = \det[G_{\alpha\beta}]_1^2 \tag{1.62}$$

If we now introduce the components (1.60a) into (1.62), then a straightforward calculation yields the formula for G in terms of the midsurface invariants. Since it will be convenient to have a distinct symbol for the square root of the ratio of the determinants of the space and surface metric tensors, we set

$$j(z) = j(x^\alpha, z) = 1 - 2zH + z^2 K \tag{1.63}$$

and the above calculation renders G in the form

$$G = g(1 - 2zH + z^2K)^2 = g[j(z)]^2 \tag{1.64}$$

We note that $j(z)$ measures the ratio of the area of the laminar element at transverse distance z to the area of the corresponding element on the midsurface.

With the covariant components given by (1.60) it can easily be seen from the determining equations (1.42) that the components of the contravariant metric tensor G^{ij} are given by

$$G^{11} = \frac{G_{22}}{G}, \quad G^{22} = \frac{G_{11}}{G}, \quad G^{12} = -\frac{G_{12}}{G} \tag{1.65a}$$

$$G^{13} = G^{23} = 0: \quad G^{33} = 1 \tag{1.65b:c}$$

By means of the permutation tensor ϵ^{ijk} whose nonvanishing components are given by (1.46), the three relations (1.65a) may be combined in the tensor form

$$G^{\alpha\beta} = \epsilon^{\alpha\lambda 3}\epsilon^{\beta\mu 3} G_{\lambda\mu} \tag{1.66}$$

Inserting formula (1.64) for G into (1.45), we see that the components $\epsilon_{\alpha\beta 3}$ and $\epsilon^{\alpha\beta 3}$ are related to the two-dimensional permutation tensors in the following manner

$$\epsilon_{\alpha\beta 3} = j(z)\,\epsilon_{\alpha\beta}, \quad \epsilon^{\alpha\beta 3} = \frac{1}{j(z)}\epsilon^{\alpha\beta} \tag{1.67a,b}$$

If we introduce expression (1.60a) for $G_{\lambda\mu}$ into (1.66) and also use (1.67b), we have

$$G^{\alpha\beta} = \frac{1}{[j(z)]^2}\epsilon^{\alpha\lambda}\epsilon^{\beta\mu}[g_{\lambda\mu}(1 - z^2K) - 2zb_{\lambda\mu}(1 - zH)] \tag{1.68}$$

in which we now employ relations (1.12) and (1.28) to eliminate the permutation tensors yielding

$$G^{\alpha\beta} = \frac{1}{[j(z)]^2}\left[g^{\alpha\beta}(1 - z^2K) - 2z(1 - zH)(2Hg^{\alpha\beta} - b^{\alpha\beta}) \right] \tag{1.69}$$

Thus, with a slight rearrangement in the above we may now replace (1.65) by the tensor relations

$$G^{\alpha\beta} = \frac{1}{[j(z)]^2}\left[g^{\alpha\beta}[1 - 4zH + z^2(4H^2 - K)] + 2zb^{\alpha\beta}(1 - zH) \right], \ \alpha,\beta = 1,2 \tag{1.70a}$$

$$G^{\alpha 3} = 0, \quad \alpha = 1,2: \quad G^{33} = 1 \tag{1.70b:c}$$

which, with (1.58a), when introduced into the defining formulae (1.43), give for the conjugate base vectors

$$\underset{\sim}{G}^{\alpha} = \frac{1}{j(z)}\,[(1 - 2zH)\,\underset{\sim}{g}^{\alpha} + zb^{\alpha}_{\lambda}\,\underset{\sim}{g}^{\lambda}], \quad \alpha = 1,2: \quad \underset{\sim}{G}^{3} = \underset{\sim}{n} \tag{1.71a:b}$$

where, in the rearrangement of (1.71a), we utilize the contravariant form of (1.61).

Relations (1.59) and (1.69) are now to be used for the computation of the Christoffel symbols (1.47). It is immediately evident that those symbols, in which the index 3 appears at least twice, vanish

$$\Gamma^{3}_{33} = 0: \quad \Gamma^{\lambda}_{33} = 0, \quad \lambda = 1,2 \tag{1.72a:b}$$

and also

$$\Gamma^{3}_{\alpha 3} = \Gamma^{3}_{3\alpha} = 0, \quad \alpha = 1,2 \tag{1.73}$$

Moreover, for the symbols with 3 in the upper position but in neither of the lower positions, we have

$$\Gamma^{3}_{\alpha\beta} = -G_{\alpha\beta,3} = zKg_{\alpha\beta} + b_{\alpha\beta}(1 - 2zH) \tag{1.74}$$

leading to the equivalent forms

$$\Gamma^{3}_{\alpha\beta} = b_{\alpha\beta} - z(2\mathrm{H}b_{\alpha\beta} - Kg_{\alpha\beta}) \tag{1.75}$$

$$= b_{\alpha\beta} - zb^{\mu}_{\alpha}\,b_{\mu\beta} \tag{1.75*}$$

wherein we have used relation (1.61). Next, for those symbols with index 3 in one of the lower positions but not in the upper position we note that

$$\Gamma^{\lambda}_{3\beta} = \frac{1}{2}\,G^{\lambda\mu}\,G_{\beta\mu,3} \tag{1.76}$$

which, when we substitute for the metric coefficients from (1.60a) and (1.70a) and perform the multiplication, gives

$$\Gamma^{\lambda}_{3\beta} = -\frac{1}{[j(z)]^2}\Big[[1 - 6zH + z^2(12H^2 + K) - 8z^3H^3]b^{\lambda}_{\beta} + z[1 - 4zH + z^2(4H^2 - K)]K\delta^{\lambda}_{\beta} + 2z[1 - 3zH + 2z^2H^2]b^{\lambda}_{\mu}b^{\mu}_{\beta}\Big] \tag{1.77}$$

The mixed form of relation (1.61) reads

$$b^{\lambda}_{\mu}b^{\mu}_{\beta} = 2Hb^{\lambda}_{\beta} - K\delta^{\lambda}_{\beta} \tag{1.78}$$

which may now be introduced into the third factor on the right of (1.77): after a rearrangement, we find

$$\Gamma^{\lambda}_{3\beta} = -\frac{1}{[j(z)]^2}(1 - 2zH + z^2K)(b^{\lambda}_{\beta} - zK\delta^{\lambda}_{\beta}) \tag{1.79}$$

Recalling the defining formula (1.63), we may cancel the factor $j(z)$ in the numerator with one of those in the denominator again using relation (1.78), we obtain the equivalent forms

$$\Gamma^{\lambda}_{3\beta} = -\frac{1}{j(z)}[b^{\lambda}_{\beta} - zK\delta^{\lambda}_{\beta}] \tag{1.80}$$

$$= -\frac{1}{j(z)}[b^{\lambda}_{\beta}(1 - 2zH) + zb^{\lambda}_{\mu}b^{\mu}_{\beta}] \tag{1.80*}$$

Lastly, for the symbols in which the index 3 does not appear, we have

$$\Gamma^{\lambda}_{\alpha\beta} = \frac{1}{2}G^{\lambda\mu}[G_{\alpha\mu,\beta} + G_{\beta\mu,\alpha} - G_{\alpha\beta,\mu}] \tag{1.81}$$

which, when we introduce the covariant metric coefficients (1.60a) into the second factor, becomes

$$\Gamma^{\lambda}_{\alpha\beta} = \frac{1}{2}G^{\lambda\mu}\Big[(1-z^2K)[g_{\alpha\mu,\beta}+g_{\beta\mu,\alpha}-g_{\alpha\beta,\mu}] - 2z(1-zH)[b_{\alpha\mu,\beta}+b_{\beta\mu,\alpha}-b_{\alpha\beta,\mu}] + z^2[(2b_{\alpha\mu}H,_{\beta} - g_{\alpha\mu}K,_{\beta}) + (2b_{\beta\mu}H,_{\alpha} - g_{\beta\mu}K,_{\alpha}) - (2b_{\alpha\beta}H,_{\mu} - g_{\alpha\beta}K,_{\mu})]\Big] \tag{1.82}$$

From the defining relation (1.36b) for the covariant derivative, it can be easily checked that

$$b_{\alpha\mu,\beta} + b_{\beta\mu,\alpha} - b_{\alpha\beta,\mu} = b_{\alpha\mu|\beta} + b_{\beta\mu|\alpha} - b_{\alpha\beta|\mu} + 2\left\{ {\rho \atop \alpha\,\beta} \right\} b_{\rho\mu} \tag{1.83a}$$

$$= b_{\alpha\mu|\beta} + 2\left\{ {\rho \atop \alpha\,\beta} \right\} b_{\rho\mu} \tag{1.83b}$$

where the latter form follows by virtue of the Codazzi relation (1.33): if we utilize (1.83b) and also the notation of (1.13a) in (1.82), we see that

$$\Gamma^{\lambda}_{\alpha\beta} = G^{\lambda\mu}\left[(1 - z^2K)[\alpha,\beta:\mu] - z(1 - zH)\left[b_{\alpha\mu|\beta} + 2\left\{ {\rho \atop \alpha\,\beta} \right\} b_{\rho\mu}\right] + \frac{1}{2}z^2[(2b_{\alpha\mu}H_{,\beta} - g_{\alpha\mu}K_{,\beta}) + (2b_{\beta\mu}H_{,\alpha} - g_{\beta\mu}K_{,\alpha}) - (2b_{\alpha\mu}H_{,\mu} - g_{\alpha\beta}K_{,\mu})]\right] \tag{1.84}$$

We now replace the first factor by the appropriate expression for the contravariant metric coefficient as given by (1.70a) and perform the multiplication: again we utilize relation (1.78) and after a considerable rearrangement we find

$$\begin{aligned}\Gamma^{\lambda}_{\alpha\beta} = {} & \left\{ {\lambda \atop \alpha\,\beta} \right\} - z\frac{1 - zH}{[j(z)]^2}\left[[1 - 4zH + z^2(4H^2 - K)]b^{\lambda}_{\alpha|\beta} - 2z(1 - zH)b^{\lambda}_{\mu}\, b^{\mu}_{\alpha|\beta}\right] \\ & + z^2\frac{1 - z^2K}{[j(z)]^2}[b^{\lambda}_{\alpha}H_{,\beta} + b^{\lambda}_{\beta}H_{,\alpha}] - \frac{1}{2}z^2\frac{[1 - 4zH + z^2(4H^2 - K)]}{[j(z)]^2}[\delta^{\lambda}_{\alpha}K_{,\beta} + \delta^{\lambda}_{\beta}K_{,\alpha}] \\ & + \frac{1}{2}z^2\frac{1}{[j(z)]^2}g_{\alpha\beta}\left[[1 - 4zH + z^2(4H^2 - K)]g^{\lambda\mu} + 2z(1 - zH)b^{\lambda\mu}\right]K_{,\mu} \\ & - z^2\frac{1}{[j(z)]^2}b_{\alpha\beta}\left[[1 - 4zH + z^2(4H^2 - K)]g^{\lambda\mu} + 2z(1 - zH)b^{\lambda\mu}\right]H_{,\mu} \\ & - z^3\frac{(1 - zH)}{[j(z)]^2}[b^{\lambda}_{\alpha}K_{,\beta} + b^{\lambda}_{\beta}K_{,\alpha} + 2K(\delta^{\lambda}_{\alpha}H_{,\beta} + \delta^{\lambda}_{\beta}H_{,\alpha})]\end{aligned} \tag{1.85}$$

having introduced the notation of (1.13b). The latter may be written in the form

$$\Gamma^{\lambda}_{\alpha\beta} = \left\{ {\lambda \atop \alpha\,\beta} \right\} + \zeta^{\lambda}_{\alpha\beta} \tag{1.86}$$

if we use the symbol $\zeta^{\lambda}_{\alpha\beta}$ to denote the "correction" terms due to the curvature.

Using relations (1.72), (1.73), (1.75), (1.80) and (1.86), we may now write the formulae for the covariant derivative in a surface related form. In particular if we consider the vector $\underset{\sim}{A}$ with component resolution

$$\underset{\sim}{A} = a_{\alpha} \underset{\sim}{G}^{\alpha} + a_3 \underset{\sim}{n} \tag{1.87}$$

then for the normal component a_3, if we note (1.72), the transverse derivative is given by

$$a_{3\|3} = a_{3,z} \tag{1.88a}$$

while introducing (1.73) and (1.80) we obtain for the laminar derivative

$$a_{3\|\beta} = a_{3,\beta} + \frac{1}{[j(z)]} [b^{\lambda}_{\beta}(1-2zH) + zb^{\lambda}_{\mu} b^{\mu}_{\beta}]a_{\lambda} \tag{1.88b}$$

Similarly for the laminar components a_{α} the transverse derivative is given by

$$a_{\alpha\|3} = a_{\alpha,z} + \frac{1}{j(z)} [b^{\lambda}_{\beta}(1-2zH) + zb^{\lambda}_{\mu} b^{\mu}_{\beta}]a_{\lambda} \tag{1.88c}$$

while from (1.75) and (1.86) there follows for the laminar derivative

$$a_{\alpha\|\beta} = a_{\alpha|\beta} - \zeta^{\lambda}_{\alpha\beta} a_{\lambda} - (b_{\alpha\beta} - zb^{\lambda}_{\alpha} b_{\lambda\beta})a_3 \tag{1.88d}$$

where we have introduced the notation of (1.35b).

For the description of the shell we first note that referred to the Gaussian coordinates $[x^{\alpha}, \alpha = 1,2]$ the equation of the edge curve along which the boundary intersects the midsurface may be written in the parametric form

$$\mathfrak{C}: \qquad x^{\alpha} = \overline{x^{\alpha}(\eta)},\ \alpha = 1,2;\ 0 \leq \eta \leq 1\ . \tag{1.89}$$

where the $\overline{x^{\alpha}}$ are given functions of the normalized running parameter η on the curve. The nonparametric form

$$\mathfrak{C}: \qquad Y(x^{\alpha}) = Y(x^1, x^2) = 0 \tag{1.90}$$

follows from the elimination of η from (1.89) and hence, the set

$$\mathfrak{A}: \qquad Y(x^{\alpha}) \leq 0 \tag{1.91}$$

describes the region occupied by the projection of the shell on the midsurface. Accordingly, if we let h denote the half thickness of the shell, then, referred to the conoidal coordinate system $[x^\alpha, z]$, the boundary consists of the ruled edge surface

$$\mathcal{B}: \qquad x^\alpha = \overline{x^\alpha(\eta)}, \ \alpha = 1,2; \ 0 \le \eta \le 1: \ -h \le z \le h \tag{1.92a}$$

together with the pair of faces

$$z = \mp h, \ Y(x^\alpha) \le 0 \tag{1.92b}$$

and hence the region

$$\mathcal{R}: \qquad Y(x^\alpha) \le 0, \ |z| \le h \tag{1.93}$$

defines the interior of the shell.

We shall use a bar to indicate when quantities are evaluated on the edge so that on the curve (1.89) we write for the base vectors

$$\overline{\underset{\sim}{g}_\alpha} = \overline{\underset{\sim}{g}_\alpha(\eta)} = \left[\underset{\sim}{g}_\alpha(x^\lambda)\right]_{\overline{x^\lambda = x^\lambda(\eta)}} \tag{1.94a}$$

$$\overline{\underset{\sim}{g}^\alpha} = \overline{\underset{\sim}{g}^\alpha(\eta)} = \left[\underset{\sim}{g}^\alpha(x^\lambda)\right]_{\overline{x^\lambda = x^\lambda(\eta)}} \tag{1.94b}$$

and for the metric and permutation tensors

$$\overline{g_{\alpha\beta}} = \overline{g_{\alpha\beta}(\eta)} = \left[g_{\alpha\beta}(x^\lambda)\right]_{\overline{x^\lambda = x^\lambda(\eta)}} \tag{1.95a}$$

$$\overline{\epsilon_{\alpha\beta}} = \overline{\epsilon_{\alpha\beta}(\eta)} = \left[\epsilon_{\alpha\beta}(g)\right]_{\overline{g = g(\eta)}} \tag{1.95b}$$

where in the latter we have noted (1.9) and set

$$\overline{g(\eta)} = \det\overline{[g_{\alpha\beta}(\eta)]}_1^2 \tag{1.96}$$

For the curve (1.89), the contravariant components of the tangent vector referred to the surface base vectors $\underset{\sim}{g}_\alpha$ are

$$\chi^\alpha(\eta) = \frac{d}{d\eta}\overline{x^\alpha(\eta)} \tag{1.97}$$

and hence, the covariant quantities

$$y_\alpha(\eta) = \overline{\epsilon_{\alpha\beta}(\eta)}\ \chi^\beta(\eta) \tag{1.98}$$

constitute the principal normal vector. If we introduce the associated normalized components $\mathfrak{n}_\alpha(\eta)$ by

$$\mathfrak{n}_\alpha(\eta) = \frac{y_\alpha(\eta)}{\sqrt{y_\lambda(\eta)\, y^\lambda(\eta)}} \tag{1.99}$$

then referred to the surface vectors $\underset{\sim}{g}_\alpha$ we have resolution

$$\underset{\sim}{\mathfrak{n}} = \mathfrak{n}_\alpha\, \overline{\underset{\sim}{g}}^\alpha \tag{1.100}$$

for the unit vector $\underset{\sim}{\mathfrak{n}}$ along the principal normal to the edge curve.

Consistent with the notation of (1.94), we use a bar to distinguish the forms $[\overline{\underset{\sim}{G}_\alpha}, \overline{\underset{\sim}{n}}]$ and $[\overline{\underset{\sim}{G}^\alpha}, \overline{\underset{\sim}{n}}]$ assumed by the space triads on the edge surface. We next obtain the representation, referred to the triad $[\overline{\underset{\sim}{G}^\alpha}, \overline{\underset{\sim}{n}}]$ of the unit normal $\underset{\sim}{\mathfrak{n}}$ to the edge surface (1.92a) at an arbitrary point. The edge surface may be generated by parallel translation of the edge curve (1.89) along the binormal $\overline{\underset{\sim}{n}}$. In the conoidal coordinate system the parallel curve at transverse distance z is given parametrically by

$$x^\alpha = \overline{x^\alpha(\eta)}, \quad \alpha = 1,2: \quad z = z \tag{1.101}$$

and hence, referred to the triad $[\underset{\sim}{G}_\alpha, \underset{\sim}{n}]$, we have for the components $\tilde{\chi}^i$ of the tangent vector

$$\tilde{\chi}^\alpha(\eta) = \frac{d}{d\eta}\overline{x^\alpha(\eta)} = \chi^\alpha(\eta), \quad \alpha = 1,2: \quad \tilde{\chi}^3 = 0 \tag{1.102}$$

which together with the transverse vector $\overline{\underset{\sim}{n}}$ determine the plane tangent to the edge surface. Recalling formula (1.47) for the cross product and also relation (1.58b) giving $\overline{\underset{\sim}{n}} = \overline{\underset{\sim}{G}_3}$, we derive the components $\tilde{y}_i$ of the vector normal to the edge surface at an arbitrary point (η, z),

$$\tilde{y}_i(\eta,z) = \overline{\epsilon_{ij3}(\eta,z)}\ \tilde{\chi}^j(\eta) \tag{1.103}$$

where

$$\overline{\epsilon_{ijk}(\eta,z)} = \left[\epsilon_{ijk}(x^{\lambda},z)\right]_{\overline{x^{\lambda}=x^{\lambda}(\eta)}} \tag{1.104}$$

We therefore have

$$\tilde{y}_{\alpha}(\eta,z) = \overline{\epsilon_{\alpha\beta 3}(\eta,z)}\; \chi^{\beta}(\eta), \quad \alpha = 1,2: \; \tilde{y}_3 = 0 \tag{1.105a,b}$$

and if we introduce relation (1.67a) and set

$$\overline{H} = \left[H(x^{\lambda})\right]_{\overline{x^{\lambda}=x^{\lambda}(\eta)}}, \quad \overline{K} = \left[K(x^{\lambda})\right]_{\overline{x^{\lambda}=x^{\lambda}(\eta)}} \tag{1.106}$$

for the edge values of the surface invariants and also write

$$\overline{j(\eta,z)} = 1 - 2z\overline{H} + z^2\overline{K} \tag{1.107}$$

then the nonvanishing components (1.105a) of the vector normal to the edge surface, may be written

$$\tilde{y}_{\alpha}(\eta,z) = \overline{j(\eta,z)}\; \overline{\epsilon_{\alpha\beta}(\eta)}\; \chi^{\beta}(\eta) = \overline{j(\eta,z)}\; y_{\alpha}(\eta) \tag{1.108}$$

where the y_{α} are the components of the principal normal to the edge curve given by (1.98). It is easily seen that the corresponding contravariant components are

$$\tilde{y}^{\alpha}(\eta,z) = \frac{1}{\overline{j(\eta,z)}}\, y^{\alpha}(\eta) \tag{1.109}$$

and hence the magnitude of the normal vector is unaffected by the transverse distance since

$$\tilde{y}_{\lambda}(\eta,z)\,\tilde{y}^{\lambda}(\eta,z) = y_{\lambda}(\eta)\, y^{\lambda}(\eta) \tag{1.110}$$

It follows from the definition of the associated normalized components, namely

$$\tilde{n}_{\alpha}(\eta,z) = \frac{\tilde{y}_{\alpha}(\eta,z)}{\sqrt{\tilde{y}_{\lambda}(\eta,z)\,\tilde{y}^{\lambda}(\eta,z)}} \tag{1.111}$$

that these may be expressed in terms of the corresponding components n_{α} of the unit principal normal to the edge curve by the relation

$$\tilde{n}_{\alpha}(\eta,z) = \overline{j(\eta,z)}\; n_{\alpha}(\eta) \tag{1.112}$$

Hence for the unit normal $\tilde{\underset{\sim}{n}}$ at an arbitrary point (η, z) of the edge surface we have the alternative resolutions

$$\tilde{\underset{\sim}{n}} = \tilde{n}_\alpha \overline{\underset{\sim}{G}^\alpha} \tag{1.113a}$$

$$= \overline{j(\eta, z)}\, n_\alpha \overline{\underset{\sim}{G}^\alpha} \tag{1.113b}$$

which at $z = 0$ clearly coincides with the unit principal normal to the edge curve given by (1.100).

On the remainder of the boundary consisting of the pair of faces (1.92b) the unit normal coincides with the unit transverse vector $\underset{\sim}{n}$ except for change of sign on the lower face.

In the sequel Greek indices will consistently have the range (1,2) while Latin indices will range over (1,2,3).

1A. The Approximation Scheme and Associated Relations

The series of relations (1.58) to (1.85) for the coordinate elements and Christoffel symbols reveal one of the sources of difficulty in the analysis of the general equations in the conoidal coordinate system. In particular we note that although the base vectors $\underset{\sim}{G}_\alpha$ (1.58a) and the covariant metric coefficients $G_{\alpha\beta}$ (1.59a) are respectively linear and quadratic in their dependence on the transverse coordinate, this relative simplicity is not shared by the conjugate system: in fact, in relations (1.70) for the contravariant metric coefficients and in formulae (1.71) for the conjuagate vectors, we see that there occur denominators which in the latter are quadratic, and in the former are biquadratic, in their dependence on z. This non-polynomial dependence propagates into the Christoffel symbols (1.80) and (1.85) thereby imposing a tedious complication on the problem of integrating the equations with respect to the transverse variable.

Even apart from this difficulty we can see from relation (1.85) that the exact integration, to eliminate the thickness coordinate from the three dimensional equations, would nevertheless pose a formidable computational problem. However if we take advantage of the characteristic features of the shell configuration, we can develop an approximate procedure that allows a relatively straightforward execution of the transverse integration.

At any point on the midsurface we have the geometrical radii of curvature R_{11} and R_{22} together with the radius of torsion R_{12} given by the relations

$$\frac{1}{R_{\alpha\beta}} = \frac{b_{\alpha\beta}}{\sqrt{g_{\alpha\alpha}\, g_{\beta\beta}}}, \quad \alpha,\beta = 1,2 \quad \text{(no summation)} \tag{1A.1}$$

In terms of these we define the critical radial measure R_* as the minimum value assumed by any of the three radii (1A.1) on the midsurface, namely

$$R_* = \min_{Y(x^\alpha)<0} \left[|R_{11}|, |R_{22}|, |R_{12}| \right] \tag{1A.2}$$

A shell configuration may be characterized by the satisfaction of the two requirements

(1) The curvature variation on the midsurface is uniformly moderate.
(2) The region is uniformly thin in the sense that the critical radius is large compared with the half-thickness.

The first requirement, excluding sharp curvature variations, may be satisfied by stipulating that the wavelengths characterizing the variation of the first and second fundamental forms be comparable, namely

$$b_{\alpha\beta,\mu} = 0\left(\frac{1}{R_*}[\alpha,\beta:\mu]\right) \tag{1A.3}$$

The second condition, which also guarantees that the conoidal coordinate system is free of singularities in the region of interest, may be formally stated by introducing θ as the thickness-curvature parameter, namely

$$\theta = \frac{h}{R_*} \tag{1A.4}$$

so that the shell characteristic is equivalent to the comparison inequality

$$\theta < < 1 \tag{1A.5}$$

The mensural feature (1A.5) admits the treatment of the transverse integration by successive approximations with the square of the thickness-curvature parameter (θ^2) playing the role of the small parameter. We shall be satisfied with a first order theory based on the neglect of effects of order θ^2 in comparison with unity. An inspection of relations (1.60) to (1.71) shows that such an approximation allows all coordinate elements to be linearized in their dependence on the transverse coordinate. These approximate formulae then yield linear representations for the Christoffel symbols which in turn lead to a form of the constitutive relations that can be integrated with the specified degree of accuracy.

The base vectors (1.59) are unaffected by the approximation so that we have as before

$$\underset{\sim}{G}_\alpha = \underset{\sim}{g}_\alpha - zb^\lambda_\alpha \underset{\sim}{g}_\lambda, \quad \alpha = 1,2; \quad \underset{\sim}{G}_3 = \underset{\sim}{n} \tag{1A.6a:b}$$

but by neglecting terms of order θ^2 from (1.60) we obtain the approximate covariant metric coefficients, namely

$$G_{\alpha\beta} = g_{\alpha\beta} - 2zb_{\alpha\beta}, \quad \alpha,\beta = 1,2 \tag{1A.7a}$$

$$G_{\alpha 3} = 0, \quad \alpha = 1,2: \quad G_{33} = 1 \tag{1A.7b:c}$$

The determinant G is still written in the form

$$G = g[j(z)]^2 \tag{1A.8}$$

where for the ratio $j(z)$ and its reciprocal we may now take

$$j(z) = 1 - 2zH, \quad \frac{1}{j(z)} = 1 + 2zH \tag{1A.9a,b}$$

respectively. From (1A.9) it follows that

$$\frac{1}{[j(z)]^2} = \frac{1}{1 - 4zH} = 1 + 4zH \tag{1A.10}$$

which is to be introduced into formulae (1.70) for the contravariant metric coefficients: by neglecting terms of order θ^2 we then obtain the approximate form

$$G^{\alpha\beta} = g^{\alpha\beta} + 2zb^{\alpha\beta}, \quad \alpha,\beta = 1,2 \tag{1A.11a}$$

$$G^{\alpha 3} = 0, \quad \alpha = 1,2: \quad G^{33} = 1 \tag{1A.11b:c}$$

A similar approximation applied to (1.71) would yield the simpler form of the conjugate base vectors

$$\underset{\sim}{G}^{\alpha} = \underset{\sim}{g}^{\alpha} + zb^{\alpha}_{\lambda}\underset{\sim}{g}^{\lambda}, \quad \alpha = 1,2: \quad \underset{\sim}{G}^{3} = \underset{\sim}{n} \tag{1A.12a:b}$$

which however, we shall not have occasion to use.

In the case of the Christoffel symbols only those given by (1.80) and (1.85) are affected by the approximation. As in (1.72), (1.73) and (1.75*) we have

$$\Gamma^3_{33} = 0: \quad \Gamma^{\lambda}_{33} = 0, \quad \lambda = 1,2 \tag{1A.13a:b}$$

$$\Gamma^3_{\alpha 3} = \Gamma^3_{3\alpha} = 0, \quad \alpha = 1,2 \tag{1A.14}$$

$$\Gamma^3_{\alpha\beta} = b_{\alpha\beta} - zb^{\lambda}_{\alpha}\, b_{\lambda\beta} \tag{1A.15}$$

whereas the introduction of (1A.9) into (1.80*), if we neglect terms of order θ^2, gives the approximate form

$$\Gamma^{\lambda}_{3\beta} = -[b^{\lambda}_{\beta} + zb^{\lambda}_{\mu}\, b^{\mu}_{\beta}] \tag{1A.16}$$

A similar use of the approximation (1A.10) in relation (1.85) together with the omission of terms of order θ^2 effects a considerable simplification in the latter formula, namely

$$\Gamma^{\lambda}_{\alpha\beta} = \left\{ \begin{matrix} \lambda \\ \alpha\ \beta \end{matrix} \right\} - z b^{\lambda}_{\alpha|\beta} \tag{1A.17}$$

which can be used in place of (1.86)

We now apply these approximations to the expressions for the covariant derivatives of the components of the vector field

$$\underset{\sim}{A} = a_{\alpha} \underset{\sim}{G}^{\alpha} + a_3 \underset{\sim}{n} \tag{1A.18}$$

For the normal component a_3 the transverse derivative (1.88a) is unaffected, namely

$$a_{3\|3} = a_{3,z} \tag{1A.19a}$$

while from (1A.14) and (1A.16) we have for the laminar derivatives

$$a_{3\|\alpha} = a_{3,\alpha} + [b^{\lambda}_{\alpha} + z b^{\lambda}_{\mu} b^{\mu}_{\alpha}] a_{\lambda} \tag{1A.19b}$$

replacing (1.88b). Similarly for the transverse derivative of the laminar components we have instead of (1.88c)

$$a_{\alpha\|3} = a_{\alpha,z} + [b^{\lambda}_{\alpha} + z b^{\lambda}_{\mu} b^{\mu}_{\alpha}] a_{\lambda} \tag{1A.19c}$$

while, using (1A.15) and (1A.17) formula (1.88d) for the laminar derivatives is approximated by

$$a_{\alpha\|\beta} = a_{\alpha|\beta} + z b^{\lambda}_{\alpha|\beta} a_{\lambda} - (b_{\alpha\beta} - z b^{\lambda}_{\alpha} b_{\lambda\beta}) a_3 \tag{1A.19d}$$

where, as before, we utilize the notation of (1.35b). These expressions will be used in obtaining approximate expressions for the strain measures.

2. The Boundary Value Problem

For a three dimensional medium referred to a general coordinate system $[x^i, i = 1,2,3]$, in which the stress tensor is represented by $\tilde{\tau}^{ij}$, the equilibrium conditions* for a material element consist of the symmetry relations

$$\tilde{\tau}^{ij} = \tilde{\tau}^{ji} \tag{2.1}$$

together with the vector equation, which in the absence of body forces takes the form

$$\frac{\partial}{\partial x^i} [\sqrt{G}\, \tilde{\tau}^{ij} \underset{\sim}{G}_j] = 0 \tag{2.2}$$

where again summation is implied over the repeated indices: in (2.1) the base

* For the derivation of these and other basic equations we refer to the standard treatises, e.g. "Theoretical Elasticity" by A. E. Green and W. Zerna.

vectors $\underset{\sim}{G}_i$ are derived from the position vector $\underset{\sim}{R}(x^i)$ in accordance with (1.39) and G is the determinant of the metric coefficients as defined in (1.40) and (1.41). The vector equation (2.2) is equivalent to the system of tensor equations

$$\tilde{\tau}^{ij}\|_i = 0 \qquad (2.3)$$

where as before the double vertical stroke indicates covariant differentiation as defined for the metric G_{ij} by relations (1.48b) and (1.53). Using formula (1.72) to (1.85), the above equations could now be written in the particular form they take in the coordinate system natural to the shell configuration.

The stress tensor $\tilde{\tau}^{ij}$ is defined with reference to the base vectors $\underset{\sim}{G}_i$ induced by the coordinate system which for the conoidal system of the shell consist of the triad $[\underset{\sim}{G}_\alpha, \underset{\sim}{n}]$. However on the edge surface the prescription of the stress vector is generally referred to the midsurface triad $[\underset{\sim}{g}_\alpha, \underset{\sim}{n}]$: this feature of the boundary conditions makes it desirable that the stress equations governing internal equilibrium be given in terms of quantities that represent the stress field as related to these midsurface base vectors. Accordingly, we shall write the equilibrium conditions in terms of the components of a pseudo-stress-tensor τ^{ij} which gives an alternative representation of the stress field, in that it is referred to the imposed triad of base vectors. This is most conveniently done by returning to the vector equations (2.2) and independently developing the system of equations that will replace (2.3).

Giving explicit preference to the third coordinate in equation (2.2) we have

$$\frac{\partial}{\partial x^\alpha}[\sqrt{G}(\tilde{\tau}^{\alpha\beta}\underset{\sim}{G}_\beta + \tilde{\tau}^{\alpha 3}\underset{\sim}{G}_3)] + \frac{\partial}{\partial x^3}[\sqrt{G}(\tilde{\tau}^{3\beta}\underset{\sim}{G}_\beta + \tilde{\tau}^{33}\underset{\sim}{G}_3)] = 0 \qquad (2.4)$$

Next we identify the first two coordinates $[x^\alpha, \alpha = 1,2]$ with the laminar system $[x^\alpha, \alpha = 1,2]$ and equate x^3 with the transverse coordinate z so that in the conoidal system, equation (2.4) becomes

$$\frac{\partial}{\partial x^\alpha}[\sqrt{G}(\tilde{\tau}^{\alpha\beta}\underset{\sim}{G}_\beta + \tilde{\tau}^{\alpha 3}\underset{\sim}{n})] + \frac{\partial}{\partial z}[\sqrt{G}(\tilde{\tau}^{3\beta}\underset{\sim}{G}_\beta + \tilde{\tau}^{33}\underset{\sim}{n})] = 0 \qquad (2.5)$$

where we have noted (1.58b). In order to transform the latter into a vector equation referred to the midsurface triad $[g_\alpha, n]$ we introduce relations (1.58a) for the $\underset{\sim}{G}_\alpha$ into (2.5): if we also note formula (1.64) for G and make one interchange of dummy indices, we obtain

$$\frac{\partial}{\partial x^\alpha}\left[j(z)\sqrt{g}\,[(\tilde{\tau}^{\alpha\beta} - z b^\beta_\lambda \tilde{\tau}^{\alpha\lambda})\underset{\sim}{g}_\beta + \tilde{\tau}^{\alpha 3}\underset{\sim}{n}]\right]$$

$$\frac{\partial}{\partial z}\left[j(z)\sqrt{g}\,[(\tilde{\tau}^{3\beta} - z b^\beta_\lambda \tilde{\tau}^{3\lambda})\underset{\sim}{g}_\beta + \tilde{\tau}^{33}\underset{\sim}{n}]\right] = 0 \qquad (2.6)$$

Equation (2.6) indicates how the components of the pseudo-stress tensor τ^{ij} are to be defined: we set

$$\tau^{\alpha\beta} = j(z)(\tilde{\tau}^{\alpha\beta} - zb^{\beta}_{\lambda}\tilde{\tau}^{\alpha\lambda}), \quad \tau^{\alpha 3} = j(z)\,\tilde{\tau}^{\alpha 3} \tag{2.7a,b}$$

$$\tau^{3\beta} = j(z)(\tilde{\tau}^{3\beta} - zb^{\beta}_{\lambda}\tilde{\tau}^{3\lambda}), \quad \tau^{33} = j(z)\,\tilde{\tau}^{33} \tag{2.7c,d}$$

so that equation (2.6) becomes

$$\frac{\partial}{\partial x^{\alpha}}[\sqrt{g}\,(\tau^{\alpha\beta}\,\underline{g}_{\beta} + \tau^{\alpha 3}\,\underline{n})] + \frac{\partial}{\partial z}[\sqrt{g}\,(\tau^{3\beta}\,\underline{g}_{\beta} + \tau^{33}\,\underline{n})] = 0 \tag{2.8}$$

We now perform the differentiation and utilize relations (1.15), (1.21) and (1.24): omitting the common factor $\sqrt{g}$ we rearrange and obtain

$$[\tau^{\alpha\beta}{}_{,\alpha} + \begin{Bmatrix}\lambda\\ \alpha\,\lambda\end{Bmatrix}\tau^{\alpha\beta} + \begin{Bmatrix}\beta\\ \lambda\,\alpha\end{Bmatrix}\tau^{\alpha\lambda} - b^{\beta}_{\alpha}\tau^{\alpha 3} + \tau^{3\beta}{}_{,z}]\,\underline{g}_{\beta}$$

$$+ [\tau^{\alpha 3}{}_{,z} + \begin{Bmatrix}\lambda\\ \alpha\,\lambda\end{Bmatrix}\tau^{\alpha 3} + b_{\alpha\beta}\tau^{\alpha\beta} + \tau^{33}{}_{,z}]\,\underline{n} = 0 \tag{2.9}$$

If we introduce the notation (1.35a) and (1.36a) into (2.9) and then write in component form, we obtain the system of differential equations for the pseudo-stress components τ^{ij}, namely

$$\tau^{\alpha\beta}|_{\alpha} - b^{\beta}_{\alpha}\tau^{\alpha 3} + \tau^{3\beta}{}_{,z} = 0 \tag{2.10a}$$

$$\tau^{\alpha 3}|_{\alpha} + b_{\alpha\beta}\tau^{\alpha\beta} + \tau^{33}{}_{,z} = 0 \tag{2.10b}$$

Furthermore it follows by direct substitution from (2.7) that

$$\tau^{\alpha\beta} - zb^{\alpha}_{\lambda}\tau^{\lambda\beta} = j[\tilde{\tau}^{\alpha\beta} - zb^{\alpha}_{\lambda}\tilde{\tau}^{\lambda\beta} - zb^{\beta}_{\lambda}\tilde{\tau}^{\alpha\lambda} + z^{2}b^{\alpha}_{\lambda}b^{\beta}_{\mu}\tilde{\tau}^{\lambda\mu}] \tag{2.11a}$$

$$\tau^{\alpha 3} - zb^{\alpha}_{\lambda}\tau^{\lambda 3} = j[\tilde{\tau}^{\alpha 3} - zb^{\alpha}_{\lambda}\tilde{\tau}^{\lambda 3}] \tag{2.11b}$$

so that the symmetry requirements (2.1) on the stress tensor $\tilde{\tau}^{ij}$, namely

$$\tilde{\tau}^{\alpha\beta} = \tilde{\tau}^{\beta\alpha}, \quad \tilde{\tau}^{\alpha 3} = \tilde{\tau}^{3\alpha} \tag{2.12a,b}$$

are, for the pseudo-stress tensor τ^{ij}, equivalent to the conditions

$$\tau^{\alpha\beta} - zb^{\alpha}_{\lambda}\tau^{\beta\lambda} = \tau^{\beta\alpha} - zb^{\beta}_{\lambda}\tau^{\alpha\lambda} \tag{2.13a}$$

$$\tau^{\alpha 3} - zb^{\alpha}_{\lambda}\tau^{\lambda 3} = \tau^{3\alpha} \tag{2.13b}$$

The differential equations (2.10) together with the symmetry conditions (2.13) constitute the full set of equilibrium conditions on the field quantities τ^{ij} to be satisfied in the region $\mathcal{R}$.

It is important to remember that the quantities τ^{ij} are strictly neither stress quantities, nor the components of a tensor in the three dimensional space. However, as they are referred to the surface vectors $\underset{\sim}{g}_\alpha$, the quantities $\tau^{3\alpha}$ are the components of a first order surface tensor: similarly the quantities $\tau^{\alpha\beta}$ constitute a corresponding second order tensor. For such surface tensors the lowering of indices is effected through the surface metric $g_{\alpha\beta}$: this is in contrast to the stress tensor $\tilde{\tau}^{ij}$ where the change to covariant representations is made through the spatial metric tensor G_{ij} as given by (1.60). As we wish to avoid the tiresome repetition of the prefix "pseudo" in indicating the quantities τ^{ij}, we shall, having emphasized the above cautionary remarks, hereafter refer to them loosely as stress quantities or stresses.

For the displacement vector field $\underset{\sim}{V}$ we have the resolution

$$\underset{\sim}{V} = v_i \, \underset{\sim}{G}^i = v_\alpha \, \underset{\sim}{G}^\alpha + v_3 \, \underset{\sim}{n} \tag{2.14}$$

and hence, utilizing formulae (1.88), it is easily shown that the linearized strain tensor

$$\gamma_{ij} = \frac{1}{2}(v_{i\|j} + v_{j\|i}) \tag{2.15}$$

consists of the system of strain measures

$$\gamma_{33} = v_{3,z} \tag{2.16a}$$

$$\gamma_{\alpha 3} = \frac{1}{2}(v_{3,\alpha} + v_{\alpha,z}) + \frac{1}{j(z)}[b^\lambda_\alpha(1 - 2zH) + zb^\lambda_\mu \, b^\mu_\alpha] \, v_\lambda \tag{2.16b}$$

$$\gamma_{\alpha\beta} = \frac{1}{2}(v_{\alpha|\beta} + v_{\beta|\alpha}) - \zeta^\lambda_{\alpha\beta} \, v_\lambda - (b_{\alpha\beta} - zb^\lambda_\alpha \, b_{\lambda\beta})v_3 \tag{2.16c}$$

where the single stroke denote covariant differentiation as defined in (1.35b). The components of infinitesimal rotation associated with the displacement vector (2.14) are the laminar rotation

$$\psi = \frac{1}{2}(v_{1\|2} - v_{2\|1}) \tag{2.17}$$

$$= \frac{1}{2}(v_{1,2} - v_{2,1}) \tag{2.17*}$$

together with the two transverse components

$$\omega_\alpha = \frac{1}{2}(v_{\alpha\|3} - v_{3\|\alpha}) \tag{2.18}$$

$$= \frac{1}{2}(v_{\alpha,z} - v_{3,\alpha}) \tag{2.18*}$$

where, in each formula, we have noted the cancellation of the terms arising from the Christoffel symbols, due to the symmetry of the latter with respect to their lower indices.

We consider a homogeneous material with linear stress-strain law, which, in the shell configuration, has the property of laminar isotropy: this latter feature signifies invariance in the mechanical characteristics under changes of direction in laminae parallel to the midsurface, but admits a distinguished response in the transverse direction. In the conoidal coordinate system, the constitutive relations for such a material take the form

$$\gamma_{33} = \frac{1}{E_z}\tilde{\tau}_{33} - \frac{\nu_*}{E}\tilde{\tau}^{\lambda}_{\lambda} \tag{2.19a}$$

$$\gamma_{\alpha 3} = \frac{1}{B}\tilde{\tau}_{\alpha 3} \tag{2.19b}$$

$$\gamma_{\alpha\beta} = \frac{(1+\nu)}{E}\tilde{\tau}_{\alpha\beta} - G_{\alpha\beta}\frac{1}{E}[\nu\,\tilde{\tau}^{\lambda}_{\lambda} + \nu_*\,\tilde{\tau}_{33}] \tag{2.19c}$$

in which the covariant components of the stress tensor appearing on the right are in accordance with (1.60), namely

$$\tilde{\tau}^{\alpha\beta} = G_{\alpha\lambda}\,G_{\beta\mu}\,\tilde{\tau}^{\lambda\mu}, \quad \tilde{\tau}_{33} = \tilde{\tau}^{3}_{3} = \tilde{\tau}^{33} \tag{2.20a,b}$$

$$\tilde{\tau}_{\alpha 3} = G_{\alpha\lambda}\,\tilde{\tau}^{\lambda}_{3} = G_{\alpha\lambda}\,\tilde{\tau}^{\lambda 3} \tag{2.20c}$$

and the invariant $\tilde{\tau}^{\lambda}_{\lambda}$ is the trace of the laminar components of the stress tensor, namely

$$\tilde{\tau}^{\lambda}_{\lambda} = G_{\lambda\mu}\,\tilde{\tau}^{\lambda\mu} \tag{2.20d}$$

In the case of complete isotropy, the Young's moduli and Poisson's ratio associated with the laminar and transverse directions coincide, namely

$$E_z = E, \quad \nu_* = \nu \tag{2.21a,b}$$

and the transverse shear modulus is related to E and ν by

$$B = \frac{E}{(1+\nu)} \tag{2.21c}$$

so that the number of independent coefficients is then reduced to two.

In our prescription of the boundary conditions, the faces are subjected to a purely normal pressure, while on the cylindrical edge we consider a general stress vector field distributed in an arbitrary manner. If we let $\tilde{p}^-$ and $\tilde{p}^+$ denote the pressures on the lower and upper faces respectively, the conditions on the lower face read

$$z = -h, \quad Y(x^{\lambda}) \le 0: \quad \tilde{\tau}^{3\alpha}(x^{\lambda}, -h) = 0, \quad \tilde{\tau}^{33}(x^{\lambda}, -h) = \tilde{p}^- \tag{2.22a,b}$$

while, on the upper face, we have

$$z = +h, \quad Y(x^{\lambda}) \le 0: \quad \tilde{\tau}^{3\alpha}(x^{\lambda}, +h) = 0, \quad \tilde{\tau}^{33}(x^{\lambda}, +h) = \tilde{p}^{+} \qquad \textbf{(2.23a,b)}$$

At an arbitrary point (η, z) of the cylindrical surface, we consider the applied stress vector $\underset{\sim}{Z}(\eta, z)$ with the resolution with respect to the midsurface triad $(\underset{\sim}{g}_{\alpha}, \underset{\sim}{n})$, given by

$$\underset{\sim}{Z} = Z^{\alpha}(\eta, z)\, \overline{\underset{\sim}{g}}_{\alpha} + Z^{3}(\eta, z)\, \overline{\underset{\sim}{n}} \qquad \textbf{(2.24)}$$

If we define quantities $\tilde{Z}^{i}(\eta, z)$ by the relations

$$Z^{\alpha} = \tilde{Z}^{\alpha} - z\overline{b_{\lambda}^{\alpha}}\, \tilde{Z}^{\lambda}, \quad Z^{3} = \tilde{Z}^{3} \qquad \textbf{(2.25)}$$

where $\overline{b_{\lambda}^{\alpha}}$ denotes the edge value of b_{α}^{λ}, then we have the alternative resolution referred to the coordinate triad $(\overline{\underset{\sim}{G}}_{\alpha}, \overline{\underset{\sim}{n}})$

$$\underset{\sim}{Z} = \tilde{Z}^{\alpha}\, \overline{\underset{\sim}{G}}_{\alpha} + \tilde{Z}^{3}\, \overline{\underset{\sim}{n}} \qquad \textbf{(2.26)}$$

for the applied stress vector and the boundary conditions on the edge surface take the form

$$-h \le z \le +h, \; x^{\lambda} = \overline{x^{\lambda}(\eta)}: \begin{cases} \overline{\tilde{\tau}^{\alpha\beta}}(\eta, z)\, \tilde{n}_{\alpha}(\eta, z) = \tilde{Z}^{\beta}(\eta, z) & \textbf{(2.27a)} \\ \overline{\tilde{\tau}^{\alpha 3}}(\eta, z)\, \tilde{n}_{\alpha}(\eta, z) = \tilde{Z}^{3}(\eta, z) & \textbf{(2.27b)} \end{cases}$$

where the $\tilde{n}_{\alpha}(\eta, z)$ denote the components of the unit normal vector (1.113) and we have written

$$\overline{\tilde{\tau}^{\alpha\beta}}(\eta, z) = \left[\tilde{\tau}^{\alpha\beta}(x^{\lambda}, z)\right]_{x^{\lambda} = \overline{x^{\lambda}(\eta)}} \qquad \textbf{(2.28a)}$$

$$\overline{\tilde{\tau}^{\alpha 3}}(\eta, z) = \left[\tilde{\tau}^{\alpha 3}(x^{\lambda}, z)\right]_{x^{\lambda} = \overline{x^{\lambda}(\eta)}} \qquad \textbf{(2.28b)}$$

to signify the edge values of the stress components.

We now transform the above boundary requirements so as to express conditions on the modified stress quantities τ^{ij}. Starting with the face conditions (2.22a) and (2.23a), we first note, that since we have assumed zero values on the faces for the shear stress $\tilde{\tau}^{3\alpha}$, the homogeneity of the transformation (2.7c) implies that the modified quantities $\tau^{3\alpha}$ also satisfy homogeneous face conditions. We next note the transformation (2.7d) for the normal stress, and accordingly, multiply (2.22b) by the factor $j(-h)$, and similarly, multiply (2.23b) by $j(+h)$: if we then set

$$p^{-} = j(-h)\,\tilde{p}^{-} = (1 + 2hH + h^{2}K)\,\tilde{p}^{-} \quad \textbf{(2.29a)}$$

$$p^{+} = j(+h)\,\tilde{p}^{+} = (1 - 2hH + h^{2}K)\,\tilde{p}^{+} \quad \textbf{(2.29b)}$$

we see that, in terms of the modified stresses, the face conditions become

$$z = -h, \quad Y(x^{\lambda}) \le 0: \quad \tau^{3\alpha}(x^{\lambda}, -h) = 0, \quad \tau^{33}(x^{\lambda}, -h) = p^{-} \quad \textbf{(2.30a,b)}$$

$$z = +h, \quad Y(x^{\lambda}) \le 0: \quad \tau^{3\alpha}(x^{\lambda}, +h) = 0, \quad \tau^{33}(x^{\lambda}, +h) = p^{+} \quad \textbf{(2.31a,b)}$$

In transforming the edge conditions, we first note relations (2.25) and multiply both sides of (2.27a) by the factor $(\delta^{\lambda}_{\alpha} - z\overline{b^{\lambda}_{\alpha}})$, so that, in terms of the original resolution (2.24) of the applied stress vector, the conditions (2.27) read

$$-h \le z \le +h, \; x^{\lambda} = \overline{x^{\lambda}(\eta)}: \begin{cases} (\overline{\tilde{\tau}^{\alpha\beta}} + z\overline{b^{\alpha}_{\lambda}}\,\overline{\tilde{\tau}^{\lambda\beta}})\,\tilde{\mathrm{n}}_{\alpha} = Z^{\beta} & \textbf{(2.32a)} \\ \overline{\tilde{\tau}^{\alpha 3}}\,\tilde{\mathrm{n}}_{\alpha} = Z^{3} & \textbf{(2.32b)} \end{cases}$$

We now use formula (1.112) for the $\tilde{\mathrm{n}}_{\alpha}$ in (2.32), and, recalling the transformations (2.7b,d), we see that, in terms of the modified stress quantities, the edge conditions take the form

$$-h \le z \le +h, \; x^{\lambda} = \overline{x^{\lambda}(\eta)}: \begin{cases} \overline{\tau^{\alpha\beta}(\eta, z)}\; \mathrm{n}_{\alpha}(\eta) = Z^{\beta}(\eta, z) & \textbf{(2.33a)} \\ \overline{\tau^{\alpha 3}(\eta, z)}\; \mathrm{n}_{\alpha}(\eta) = Z^{3}(\eta, z) & \textbf{(2.33b)} \end{cases}$$

where the n_{α} are the components of the principal normal to the edge curve as given by (1.100).

The equilibrium equations (2.10) and (2.13), when combined with formula (2.7) are complemented by the constitutive relations (2.19) with the associated strain-displacement relations (2.16): the boundary value problem consists of the solution of this system of relations in the region $\mathcal{R}$ subject to the boundary conditions (2.30), (2.31) and (2.33).

2A. The Approximate Constitutive Relations

For a more consistent formulation, it is desirable that the constitutive relations be transformed so as to relate the strain measures to the modified stress quantities τ^{ij}. In the transformed system, the factor $j(z)$ would appear as a denominator in the right hand side, so that this is an appropriate point for the introduction of the approximation scheme outlined in Section 1A. In accordance with the procedure for a first order theory, we neglect terms of order θ^2 in these expressions involving the modified stresses. It is, therefore, consistent to make a similar approximation in the expressions for the strain measures appearing on the left. We are thus led to a system of stress-displacement relations accurate to first order in the small parameter θ^2.

In considering the right hand side of the first relation (2.19a), we first note from (2.20b) and (2.7d) that

$$\tilde{\tau}_{33} = \tilde{\tau}^{33} = \frac{1}{j(z)} \tau^{33} \tag{2A.1}$$

$$= (1 + 2zH)\, \tau^{33} + O(\theta^2) \tag{2A.1*}$$

while by utilizing formula (1.59a) for $G_{\lambda\mu}$ in relation (2.20d), we obtain for the invariant $\tilde{\tau}^{\lambda}_{\lambda}$,

$$\tilde{\tau}^{\lambda}_{\lambda} = [g_{\lambda\mu} - 2zb_{\lambda\mu} + z^2 b^{\rho}_{\lambda} b_{\rho\mu}]\, \tilde{\tau}^{\lambda\mu} \tag{2A.2}$$

which may be rewritten as

$$\tilde{\tau}^{\lambda}_{\lambda} = [g_{\lambda\mu} - zb_{\lambda\mu})(\tilde{\tau}^{\lambda\mu} - zb^{\lambda}_{\rho}\, \tilde{\tau}^{\mu\rho}) \tag{2A.3}$$

On the introduction of (2.7a) the latter relation becomes

$$\tilde{\tau}^{\lambda}_{\lambda} = \frac{1}{j(z)} (g_{\lambda\mu} - zb_{\lambda\mu})\, \tau^{\lambda\mu}$$

$$= \frac{1}{j(z)} g_{\lambda\mu} (\tau^{\lambda\mu} - zb^{\lambda}_{\rho}\, \tau^{\rho\mu})$$

$$= \frac{1}{j(z)} (\tau^{\lambda}_{\lambda} - zb^{\lambda}_{\mu}\, \tau^{\mu}_{\lambda}) \tag{2A.5}$$

so that from the approximation (1A.9b) we have

$$\tilde{\tau}^{\lambda}_{\lambda} = (1 + 2zH)(\tau^{\lambda}_{\lambda} - zb^{\lambda}_{\mu}\, \tau^{\mu}_{\lambda}) + O(\theta^2) \tag{2A.6}$$

$$= (1 + 2zH)\tau^{\lambda}_{\lambda} - zb^{\lambda}_{\mu}\, \tau^{\mu}_{\lambda} + O(\theta^2) \tag{2A.6*}$$

For the transverse shear stresses appearing in the second relation (2.19b), when we introduce (1.59a) into (2.20c) and note the symmetry relation (2.12b) we find

$$\tilde{\tau}_{\alpha 3} = (g_{\alpha\lambda} - 2zb_{\alpha\lambda} + z^2 b^{\mu}_{\alpha} b_{\mu\lambda}) \tilde{\tau}^{3\lambda}$$

$$= (g_{\alpha\lambda} - zb_{\alpha\lambda})(\tilde{\tau}^{3\lambda} - zb^{\lambda}_{\mu} \tilde{\tau}^{3\mu}) \qquad \textbf{(2A.7)}$$

Substituting from (2.7c) into the latter expression we obtain

$$\tilde{\tau}_{\alpha 3} = \frac{1}{j(z)} (g_{\alpha\lambda} - zb_{\alpha\lambda})\tau^{3\lambda}$$

$$= \frac{1}{j(z)} (\tau_{3\alpha} - zb^{\lambda}_{\alpha} \tau_{3\lambda}) \qquad \textbf{(2A.9)}$$

so that using the approximation (1A.9b) we have

$$\tilde{\tau}_{\alpha 3} = (1 + 2zH)(\tau_{3\alpha} - zb^{\lambda}_{\alpha} \tau_{3\lambda}) + O(\theta^2) \qquad \textbf{(2A.10)}$$

$$= (1 + 2zH)\,\tau_{3\alpha} - zb^{\lambda}_{\alpha} \tau_{3\lambda} + O(\theta^2) \qquad \textbf{(2A.10*)}$$

The covariant laminar tensor $\tilde{\tau}_{\alpha\beta}$ appearing in (2.19c) is given in (2.20a): if we insert formula (1.59a) for $G_{\beta\mu}$ we have

$$\tilde{\tau}_{\alpha\beta} = G_{\alpha\lambda}(g_{\beta\mu} - 2zb_{\beta\mu} + z^2 b_{\beta\rho} b^{\rho}_{\mu})\, \tilde{\tau}^{\lambda\mu} \qquad \textbf{(2A.11a)}$$

$$= G_{\alpha\lambda}(g_{\beta\mu} - zb_{\beta\mu})(\tilde{\tau}^{\lambda\mu} - zb^{\mu}_{\rho}\, \tilde{\tau}^{\lambda\rho}) \qquad \textbf{(2A.11b)}$$

which when we introduce (2.7a), reads

$$\tilde{\tau}_{\alpha\beta} = \frac{1}{j(z)} G_{\alpha\lambda}(g_{\beta\mu} - zb_{\beta\mu})\, \tau^{\lambda\mu} \qquad \textbf{(2A.12)}$$

By utilizing the approximation (1A.7a) and (1A.9b) in (2A.12), we obtain

$$\tilde{\tau}_{\alpha\beta} = (1 + 2zH)(g_{\alpha\lambda} - 2zb_{\alpha\lambda})(g_{\beta\mu} - zb_{\beta\mu})\, \tau^{\lambda\mu} + O(\theta^2) \qquad \textbf{(2A.13)}$$

$$= (1 + 2zH)[g_{\alpha\lambda}\, g_{\beta\mu} - z(g_{\alpha\lambda}\, b_{\beta\mu} + 2g_{\beta\mu}\, b_{\alpha\lambda})\, \tau^{\lambda\mu}] + O(\theta^2) \qquad \textbf{(2A.13*)}$$

Hence, letting the surface metric lower the indices in the surface tensor $\tau^{\lambda\mu}$, we have

$$\tilde{\tau}_{\alpha\beta} = (1 + 2zH)\,[\tau_{\alpha\beta} - z(2b_{\alpha\lambda}\, \tau^{\lambda}_{\beta} + b_{\beta\lambda}\, \tau^{\lambda}_{\beta})] + O(\theta^2) \qquad \textbf{(2A.14)}$$

$$= (1 + 2zH)\, \tau_{\alpha\beta} - z(2b_{\alpha\lambda}\, \tau^{\lambda}_{\beta} + b_{\beta\lambda}\, \tau^{\lambda}_{\alpha}) + O(\theta^2) \qquad \textbf{(2A.14*)}$$

If we had substituted for $G_{\alpha\lambda}$ rather than $G_{\beta\mu}$ in (2A.11a) and followed a similar procedure, we would have found

$$\tilde{\tau}_{\alpha\beta} = (1 + 2zH)\,\tau_{\beta\alpha} - z(b_{\alpha\lambda}\,\tau^{\lambda}_{\beta} + 2b_{\beta\lambda}\,\tau^{\lambda}_{\alpha}) + O(\theta^2) \tag{2A.15}$$

By taking the arithmetic mean of (2.14*) and (2.15) we obtain the symmetric form

$$\tilde{\tau}_{\alpha\beta} = \frac{1}{2}(1 + 2zH)(\tau_{\alpha\beta} + \tau_{\beta\alpha}) - \frac{3}{2}z(b_{\alpha\lambda}\,\tau^{\lambda}_{\beta} + b_{\beta\lambda}\,\tau^{\lambda}_{\alpha}) + O(\theta^2) \tag{2A.16}$$

In spite of the attraction of the symmetric form (2.16), we shall rather use the original form (2.14*), the advantage of which will be evident presently.

We next consider the simplification effected in the strain measures by applying the approximation scheme to expressions (2.16). The first relation (2.16a) for the transverse extensibility is unaffected, namely

$$\gamma_{33} = v_{3,z} \tag{2A.17}$$

whereas, by noting the approximations (1A.9), we see that relations (2.16b) for the transverse shear strains may be replaced by

$$2\gamma_{\alpha 3} = v_{3,\alpha} + v_{\alpha,z} + 2b^{\lambda}_{\alpha}\,v_{\lambda} + 2zb^{\lambda}_{\mu}\,b^{\mu}_{\alpha}\,v_{\lambda} + O(\theta^2) \tag{2A.18}$$

Moreover, observing that

$$v_{\alpha,z} + 2zb^{\lambda}_{\mu}\,b^{\mu}_{\alpha}\,v_{\lambda} = (v_{\alpha} + z^2\,b^{\lambda}_{\mu}\,b^{\mu}_{\alpha}\,v_{\lambda})_{,z} - z^2\,b^{\lambda}_{\mu}\,b^{\mu}_{\alpha}\,v_{\lambda,z} \tag{2A.19}$$

it is consistent with neglecting effects of order θ^2 to set

$$v_{\alpha,z} + 2zb^{\lambda}_{\mu}\,b^{\mu}_{\alpha}\,v_{\lambda} = v_{\alpha,z} \tag{2A.20}$$

and hence (2A.18) may be replaced by the approximate form

$$2\gamma_{\alpha 3} = v_{3,\alpha} + v_{\alpha,z} + 2b^{\lambda}_{\alpha}\,v_{\lambda} \tag{2A.21}$$

In obtaining the third set of measures (2.16c) we utilized relation (1.88d) for the covariant derivatives: if we neglect terms of order θ^2 we may use formula (1A.19d) in place of (1.88d) in this latter derivation and obtain the approximate form

$$\gamma_{\alpha\beta} = \frac{1}{2}(v_{\alpha|\beta} + v_{\beta|\alpha}) + zb^{\lambda}_{\alpha|\beta}\,v_{\lambda} - (b_{\alpha\beta} - zb^{\lambda}_{\alpha}\,b_{\lambda\beta})v_3 \tag{2A.22}$$

for the laminar strains.

We now combine the above relations to obtain the approximate stress-displacement relations. In (2.19a) we use relation (2A.17) on the left and introduce formulae (2A.1*) and (2A.6*) on the right: after neglecting terms of order θ^2 we have

$$v_{3,z} = (1 + 2zH)\left[\frac{1}{E_z}\tau^{33} - \frac{\nu_*}{E}\tau^{\lambda}_{\lambda}\right] + \frac{\nu_*}{E}zb^{\lambda}_{\mu}\,\tau^{\mu}_{\lambda} \tag{2A.23a}$$

Using the approximate form (2A.20) on the left of (2.19b) we insert formula (2A.10*) on the right: if we neglect effects of order θ^2, we are left with

$$v_{3,\alpha} + v_{\alpha,z} + 2b^{\lambda}_{\alpha} v_{\lambda} = \frac{2}{B}[(1 + 2zH)\,\tau_{3\alpha} - zb^{\lambda}_{\alpha}\tau_{3\lambda}] \qquad \textbf{(2A.23b)}$$

For the third stress-displacement relation we substitute the approximate form (2A.22) into the left of (2.19c) and then introduce formulae (2A.1*), (2A.6*) and (2A.16) into the right: if we also note the approximation (1A.7a) for $G_{\alpha\beta}$, then the omission of terms of order θ^2 yields after some rearrangement

$$\frac{1}{2}(v_{\alpha|\beta} + v_{\beta|\alpha}) + zb^{\lambda}_{\alpha|\beta} v_{\lambda} - (b_{\alpha\beta} - zb^{\lambda}_{\alpha} b_{\lambda\beta})v_3$$

$$= \frac{1}{E}(1 + 2zH)[(1+\nu)\,\tau_{\alpha\beta} - g_{\alpha\beta}(\nu\tau^{\lambda}_{\lambda} + \nu_* \tau^{33})]$$

$$- \frac{1}{E} z[(1 + \nu)(2b_{\alpha\lambda}\tau^{\lambda}_{\beta} + b_{\beta\lambda}\tau^{\lambda}_{\alpha}) - g_{\alpha\beta}\,\nu b^{\lambda}_{\mu}\tau^{\mu}_{\lambda} - 2b_{\alpha\beta}(\nu\tau^{\lambda}_{\lambda} + \nu_* \tau^{33})] \qquad \textbf{(2A.23c)}$$

The advantage of using the unsymmetric form (2A.14*) now becomes evident from an inspection of (2A.23c). The symmetric form of the strain measures on the left implies symmetry for the right hand side: this yields symmetry for the quantity $[(1 + 2zH)\,\tau_{\alpha\beta} - zb_{\alpha\lambda}\tau^{\lambda}_{\beta}]$ which, consistent with neglecting effects of order θ^2, is equivalent to the condition

$$\tau_{\alpha\beta} - zb_{\alpha\lambda}\tau^{\lambda}_{\beta} = \tau_{\beta\alpha} - zb_{\beta\lambda}\tau^{\lambda}_{\alpha} \qquad \textbf{(2A.24)}$$

This is the covariant form of condition (2.13a). Hence, when we consider the stress-displacement relations, the use of the unsymmetric form in the right of (2A.23c) implies conditions (2.13a) making redundant the explicit imposition of the latter. However, in our analysis of the equilibrium equations we shall, for the present, treat (2.13a) as an independent condition.

In the sequel we shall confine our consideration to the boundary value problem consisting of the exact equilibrium equations (2.10) and (2.13) together with the approximate stress-displacement relations (2A.23) and subject to the boundary conditions (2.30), (2.31) and (2.33).

3. The Normalized Formulation

To the boundary problem consisting of the exact equilibrium equations and boundary conditions complemented by the approximate constitutive relations, we now give a normalized formulation by introducing the dimensionless thickness variable t defined by

$$t = \frac{z}{h} \tag{3.1}$$

As the thickness h is constant we have

$$\frac{\partial t}{\partial x^\alpha} = 0, \quad \frac{\partial t}{\partial z} = \frac{1}{h} \tag{3.2a,b}$$

so that only the z-differentiation is affected by the transformation. The interior of the shell $\mathcal{R}$ as given in (1.93) corresponds in the transformed coordinate system (x^α, t) to the region

$$\mathcal{R}_0\colon \ -1 \le t \le +1, \quad Y(x^\alpha) \le 0 \tag{3.3}$$

whose boundary is now described by the pair of faces

$$t = \mp 1, \quad Y(x^\alpha) \le 0 \tag{3.4}$$

together with the edge surface

$$-1 \le t + 1, \; x^\alpha = \overline{x^\alpha(\eta)} \tag{3.5}$$

We now introduce notation reflecting the effect of the transformation (3.1) on the field quantities. In terms of the new coordinates, we write for the laminar stress quantities

$$\sigma^{\alpha\beta}(x^\lambda, t) = \tau^{\alpha\beta}(x^\lambda, z) \tag{3.6a}$$

and for the transverse normal stress we set

$$\sigma(x^\lambda, t) = \tau^{33}(x^\lambda, z) \tag{3.6b}$$

In transforming the transverse shear stress quantities, we distinguish between the shear stress

$$\tau^\alpha(x^\lambda, t) = \tau^{\alpha 3}(x^\lambda, z) \tag{3.6c}$$

and what we shall call the reciprocal stress

$$\sigma^\alpha(x^\lambda, t) = \tau^{3\alpha}(x^\lambda, z) \tag{3.6d}$$

In the transformed displacement quantities, we separate the transverse component from the laminar components by setting

$$u_\alpha(x^\lambda, t) = v_\alpha(x^\lambda, z), \quad \alpha = 1,2\colon \quad w(x^\lambda, t) = v_3(x^\lambda, z) \tag{3.7a:b}$$

For the relevant edge values, we set

$$\overline{\sigma^{\alpha\beta}(\eta,t)} = \left[\sigma^{\alpha\beta}(x^{\lambda},t)\right]_{\overline{x^{\lambda}=x^{\lambda}(\eta)}} = \overline{\tau^{\alpha\beta}(\eta,z)} \tag{3.8a}$$

$$\overline{\tau^{\alpha}(\eta,t)} = \left[\tau^{\alpha}(x^{\lambda},t)\right]_{\overline{x^{\lambda}=x^{\lambda}(\eta)}} = \overline{\tau^{\alpha 3}(\eta,z)} \tag{3.8b}$$

and for the associated applied edge stress vector, we write

$$T^{\alpha}(\eta,t) = Z^{\alpha}(\eta,z), \quad T(\eta,t) = Z^{3}(\eta,z) \tag{3.9a,b}$$

completing the notation necessary for the statement of the transformed boundary value problem.

In the above notation and with a slight change in the indices, the differential equations of equilibrium (2.10) become

$$h\sigma^{\beta\alpha}|_{\beta} - hb^{\alpha}_{\beta}\tau^{\beta} + \sigma^{\alpha},_{t} = 0 \tag{3.10a}$$

$$h\tau^{\alpha}|_{\alpha} + hb^{\alpha}_{\beta}\sigma^{\beta}_{\alpha} + \sigma,_{t} = 0 \tag{3.10b}$$

while the symmetry conditions (2.13) reads

$$\sigma^{\alpha\beta} - thb^{\alpha}_{\lambda}\sigma^{\lambda\beta} = \sigma^{\beta\alpha} - thb^{\beta}_{\lambda}\sigma^{\lambda\alpha} \tag{3.11a}$$

$$\tau^{\alpha} - thb^{\alpha}_{\lambda}\tau^{\lambda} = \sigma^{\alpha} \tag{3.11b}$$

Moreover, the approximate stress-displacement relations take the form

$$w,_{t} = \frac{h}{E_z}(1 + 2hHt)\,\sigma - \nu_{*}\frac{h}{E}\left[(1 + 2hHt)\,\sigma^{\lambda}_{\lambda} - thb^{\lambda}_{\mu}\sigma^{\mu}_{\lambda}\right] \tag{3.12a}$$

$$u_{\alpha,t} + 2hb^{\lambda}_{\alpha}u_{\lambda} + hw,_{\alpha} = 2\frac{h}{B}\left[(1 + 2hHt)\,\sigma_{\alpha} - thb^{\lambda}_{\alpha}\sigma_{\lambda}\right] \tag{3.12b}$$

$$\begin{aligned}\frac{1}{2}(u_{\alpha|\beta} + u_{\beta|\alpha}) &+ thb^{\lambda}_{\alpha|\beta}u_{\lambda} - (b_{\alpha\beta} - thb^{\lambda}_{\alpha}b_{\lambda\beta})w \\ &= \frac{1}{E}(1 + 2zHt)\left[(1+\nu)\,\sigma_{\alpha\beta} - g_{\alpha\beta}(\nu\sigma^{\lambda}_{\lambda} + \nu_{*}\sigma)\right] \\ &\quad - \frac{h}{E}t\left[(1+\nu)(2b_{\alpha\lambda}\sigma^{\lambda}_{\beta} + b_{\beta\lambda}\sigma^{\lambda}_{\alpha}) - g_{\alpha\beta}\nu b^{\lambda}_{\mu}\sigma^{\mu}_{\lambda} - 2b_{\alpha\beta}(\nu\sigma^{\lambda}_{\lambda} + \nu_{*}\sigma)\right]\end{aligned} \tag{3.12c}$$

complementing equations (3.10) and (3.11). For the laminar and transverse rotations, we write

$$\psi = \frac{1}{2}(u_{1,2} - u_{2,1}) \qquad \textbf{(3.13a)}$$

$$\omega_\alpha = \frac{1}{2h}(u_{\alpha,t} - hw,_\alpha) \qquad \textbf{(3.13b)}$$

respectively replacing formulae (2.17*) and (2.18*).

Writing the boundary conditions (2.30), (2.31) and (2.33) in the transformed notation, we have

$$t = -1, \quad Y(x^\lambda) \le 0: \quad \sigma^\alpha(x^\lambda, -1) = 0, \quad \sigma(x^\lambda, -1) = p^- \qquad \textbf{(3.14a,b)}$$

and

$$t = +1, \quad Y(x^\lambda) \le 0: \quad \sigma^\alpha(x^\lambda, +1) = 0, \quad \sigma(x^\lambda, +1) = p^+ \qquad \textbf{(3.15a,b)}$$

respectively for the conditions on the lower and upper faces, while the set of constraints

$$-1 \le t \le +1, \; x^\lambda = \overline{x^\lambda(\eta)}: \begin{cases} \overline{\sigma^{\alpha\beta}(\eta,t)}\, n_\beta(\eta) = T^\alpha(\eta,t) & \textbf{(3.16a)} \\ \overline{\tau^\alpha(\eta,t)}\, n_\alpha(\eta) = T(\eta,t) & \textbf{(3.16b)} \end{cases}$$

express the conditions on the edge surface.

The boundary value problem now consists of the determination of the solution of the system of equations (3.10) to (3.12), valid in the region $\mathcal{R}_0$ of (3.3) and satisfying the conditions (3.14) to (3.16).

4. Stress Representations: The Face Conditions

The reformulation of the problem in two dimensional terms requires, at the outset, that we integrate the differential equations of equilibrium (3.10) with respect to the thickness coordinate. This integration is facilitated through the expansion of the stress quantities in terms of Legendre polynomials, in which the Legendre functions reflect the transverse variation as separated from the implied dependence of the coefficients on the laminar coordinates. The symmetry conditions (3.11) yield the system of relations that must hold between the coefficients in the respective expansions while the application of the face conditions leads to the equations to be satisfied by the lower coefficients.

We start by considering the Legendre series expansion for the laminar stress tensor $\sigma^{\alpha\beta}$ together with the associated representations for the transverse shear and reciprocal stresses τ^{α} and σ^{α}. By utilizing a recursion formula for Legendre functions in the symmetry relation (3.11a) we obtain the symmetry requirements on the laminar stress coefficients. A similar procedure in the symmetry condition (3.11b) leads to a sequence of formulae expressing the elements appearing in the representation for the reciprocal shear stress σ^{α} in terms of the coefficients in the expansions for the transverse shear stresses τ^{α}.

We next introduce the series for the laminar stresses $\sigma^{\alpha\beta}$ and for the shear stresses τ^{α} into the first equilibrium differential equation (3.10a) and by means of a second recursion relation for Legendre polynomials we integrate and obtain the implied Legendre representation for the reciprocal stress σ^{α}. The form for the coefficients in the expansion of σ^{α} thus obtained, when combined with the expressions already derived from the second symmetry condition, yields the sequence of relations connecting the coefficients in the representation for the transverse shear stresses with those in the original expansion for the laminar stress tensor. By repeating the integration procedure in the second equilibrium differential equation (3.10b), we obtain the corresponding Legendre series for the transverse normal stress.

The two successive integrations introduce in turn an undetermined first order tensor and an unknown function. The satisfaction of the conditions on the upper and lower faces leads to a set of four relations involving these latter quantities together with the first two coefficients in the initial expansion. By appropriate combination of these relations, we obtain

(1) a system of coupled partial differential equations involving the zero-th and first laminar stress coefficients together with the zero-th transverse shear stress coefficient.

(2) an expression for the zero-th transverse normal stress coefficient.

The first system can be interpreted as the well-known integrated equilibrium equations of shell theory.

If we denote the Legendre polynomials of degree n in t by $P_n(t)$, the polynomials resulting from multiplication by t can be expressed in the form*

$$tP_n(t) = \frac{n+1}{2n+1} P_{n+1}(t) + \frac{n}{2n+1} P_{n-1}(t) \tag{4.1}$$

Employing a dot to signify d/dt, we have for the difference between differentiated polynomials

$$(2n+1) P_n(t) = \dot{P}_{n+1}(t) - \dot{P}_{n-1}(t) \tag{4.2}$$

which in its integrated form, apart from an additive term independent of t, reads

$$\int P_n(t)\, dt = \frac{1}{2n+1} [P_{n+1}(t) - P_{n-1}(t)] \tag{4.3}$$

* For the derivation of the recursion formulae (4.1), (4.2) and (6.1) for Legendre polynomials, we refer to § 15.21 of "Modern Analysis" by E. T. Whittaker and G. N. Watson, C.U.P., Cambridge, 1952.

Relation (4.1) will be utilized in the symmetry relations (3.11) and formula (4.3) will effect the integration of the differential equations (3.10).

We write the expansion for the laminar stress tensor in the form

$$\sigma^{\alpha\beta}(x^\lambda, t) = \sum_{n=0}^{\infty} \overset{(n)}{S}{}^{\alpha\beta}(x^\lambda)\, P_n(t) \tag{4.4}$$

in which the coefficients are undetermined functions of x^λ. Considering the left hand side of (3.11a), we introduce the expansion (4.4); and by means of the recursion formula (4.1), we obtain the Legendre series representation which, after a rearrangement of terms, reads

$$\sigma^{\alpha\beta} - thb^\alpha_\lambda \sigma^{\lambda\beta} = [\, \overset{(0)}{S}{}^{\alpha\beta} - \tfrac{1}{3} hb^\alpha_\lambda \overset{(1)}{S}{}^{\lambda\beta}]$$

$$+ \sum_{n=1}^{\infty} [\, \overset{(n)}{S}{}^{\alpha\beta} - hb^\alpha_\lambda (\frac{n}{2n-1} \overset{(n-1)}{S}{}^{\lambda\beta} + \frac{n+1}{2n+3} \overset{(n+1)}{S}{}^{\lambda\beta})]\, P_n \tag{4.5}$$

where we no longer exhibit the explicit dependence on the coordinates. When the expansion (4.5) is substituted into (3.11a), the orthogonality of the Legendre polynomials requires that the symmetry condition must be satisfied for each term independently: we are thus led to a sequence of symmetry conditions on the coefficients $\overset{(n)}{S}{}^{\alpha\beta}$, namely

$$\overset{(0)}{S}{}^{\alpha\beta} - \tfrac{1}{3} hb^\alpha_\lambda \overset{(1)}{S}{}^{\lambda\beta} = \overset{(0)}{S}{}^{\beta\alpha} - \tfrac{1}{3} hb^\beta_\lambda \overset{(0)}{S}{}^{\lambda\alpha} \tag{4.6a}$$

$$\overset{(n)}{S}{}^{\alpha\beta} - hb^\alpha_\lambda (\frac{n}{2n-1} \overset{(n-1)}{S}{}^{\lambda\beta} + \frac{n+1}{2n+3} \overset{(n+1)}{S}{}^{\lambda\beta})$$

$$= \overset{(n)}{S}{}^{\beta\alpha} - hb^\beta_\lambda (\frac{n}{2n-1} \overset{(n-1)}{S}{}^{\lambda\alpha} + \frac{n+1}{2n+3} \overset{(n+1)}{S}{}^{\lambda\alpha}),\ n \geq 1 \tag{4.6b}$$

In a similiar manner when we consider representations for the transverse shear and reciprocal stresses in the form of the Legendre series

$$\tau^\alpha = \sum_{n=0}^{\infty} \overset{(n)}{\tau}{}^\alpha(x^\lambda)\, P_n(t), \quad \sigma^\alpha = \sum_{n=0}^{\infty} \overset{(n)}{S}{}^\alpha(x^\lambda)\, P_n(t) \tag{4.7a,b}$$

we see that the symmetry relation (3.11b) is equivalent to the following sequence of relations between the coefficients

$$\overset{(0)}{S}{}^\alpha = \overset{(0)}{\tau}{}^\alpha - \tfrac{1}{3} hb^\alpha_\beta \overset{(1)}{\tau}{}^\beta \tag{4.8a}$$

$$\overset{(n)}{S}{}^{\alpha} = \overset{(n)}{\tau}{}^{\alpha\beta} - hb^{\alpha}_{\beta}\left(\frac{n}{2n-1}\overset{(n-1)}{\tau}{}^{\beta} + \frac{n+1}{2n+3}\overset{(n+1)}{\tau}{}^{\beta}\right), \; n \geq 1 \qquad \textbf{(4.8b)}$$

These will be used to relate the shear stress coefficients to the laminar stress coefficients.

In the integration of the equilibrium equations, we first introduce the expansions (4.4) and (4.7a) into equation (3.10a): after a transposition, we have

$$\sigma^{\alpha}{}_{,t} = -h\sum_{n=0}^{\infty}\overset{(n)}{S}{}^{\beta\alpha}|_{\beta}P_n + hb^{\alpha}_{\beta}\sum_{n=0}^{\infty}\overset{(n)}{\tau}{}^{\beta}P_n \qquad \textbf{(4.9)}$$

and the integrated form follows from the application of (4.3). After a rearrangement, we may write the resulting series in the form (4.7b) in which the $\overset{(0)}{S}{}^{\alpha}(x^{\lambda})$ are the components of an undetermined first order tensor introduced by the integration and the higher elements are related to the laminar and transverse shear stress coefficients by the formulae

$$\overset{(n)}{S}{}^{\alpha} = h\left[\frac{\overset{(n+1)}{S}{}^{\beta\alpha}|_{\beta}}{2n+3} - \frac{\overset{(n-1)}{S}{}^{\beta\alpha}|_{\beta}}{2n-1}\right] - hb^{\alpha}_{\beta}\left[\frac{\overset{(n+1)}{\tau}{}^{\beta}}{2n+3} - \frac{\overset{(n-1)}{\tau}{}^{\beta}}{2n-1}\right] . \; n \geq 1 \qquad \textbf{(4.10)}$$

By identifying the two alternative expressions (4.8b) and (4.10) for the higher elements in the expansion for the reciprocal stress, we obtain the sequence of formulae relating the higher shear stress coefficients to the constituents in the original expansion for the laminar stresses, namely

$$\overset{(n)}{\tau}{}^{\alpha} = h\left[\frac{\overset{(n+1)}{S}{}^{\beta\alpha}|_{\beta}}{2n+3} - \frac{\overset{(n-1)}{S}{}^{\beta\alpha}|_{\beta}}{2n-1}\right] + hb^{\alpha}_{\beta}\left[\frac{n}{2n+3}\overset{(n+1)}{\tau}{}^{\beta} + \frac{n+1}{2n-1}\overset{(n-1)}{\tau}{}^{\beta}\right], \; n \geq 1 \qquad \textbf{(4.11)}$$

The second differential equation is treated in an identical manner. After substituting the series representations (4.4) and (4.7a) into (3.10b), we transpose and obtain

$$\sigma_{,t} = -h\sum_{n=0}^{\infty}\overset{(n)}{\tau}{}^{\alpha}|_{\alpha}P_n - hb^{\alpha}_{\beta}\sum_{n=0}^{\infty}\overset{(n)}{S}{}^{\beta}{}_{\alpha}P_n \qquad \textbf{(4.12)}$$

the integration of which is also effected by means of formula (4.3). Rearranging we see that the Legendre series for the transverse normal stress may be written in the form

$$\sigma = \sum_{n=0}^{\infty}\overset{(n)}{\sigma}(x^{\lambda})P_n(t) \qquad \textbf{(4.13)}$$

in which σ_0 is undetermined and the higher terms are related to the laminar and transverse shear stress coefficients through the relations

$$\overset{(n)}{\sigma} = h\left[\frac{\overset{(n+1)}{\tau^{\alpha}}|_{\alpha}}{2n+3} - \frac{\overset{(n-1)}{\tau^{\alpha}}|_{\alpha}}{2n-1}\right] + hb^{\alpha}_{\beta}\left[\frac{\overset{(n+1)}{S^{\beta}_{\alpha}}}{2n+3} - \frac{\overset{(n-1)}{S^{\beta}_{\alpha}}}{2n-1}\right], \quad n \geq 1 \tag{4.14}$$

Before applying the face conditions, we recall the terminal values of the Legendre polynomials, namely

$$P_n(-1) = (-1)^n, \quad P_n(1) = 1 \tag{4.15}$$

which, when used in the expansions (4.7b) and (4.13), give for the face values of the reciprocal and transverse normal stress, respectively

$$\sigma^{\alpha}(x^{\lambda}, -1) = \overset{(0)}{S^{\alpha}} + \sum_{n=1}^{\infty} (-1)^n \overset{(n)}{S^{\alpha}}, \qquad \sigma^{\alpha}(x^{\lambda}, +1) = \overset{(0)}{S^{\alpha}} + \sum_{n=1}^{\infty} \overset{(n)}{S^{\alpha}} \tag{4.16a,b}$$

$$\sigma(x^{\lambda}, -1) = \overset{(0)}{\sigma} + \sum_{n=1}^{\infty} (-1)^n \overset{(n)}{\sigma}, \qquad \sigma(x^{\lambda}, +1) = \overset{(0)}{\sigma} + \sum_{n=1}^{\infty} \overset{(n)}{\sigma} \tag{4.17a,b}$$

If we insert formulae (4.10) for the $\overset{(n)}{S^{\alpha}}$ into (4.16) and note the cancellations resulting from the recursive features of (4.10), we find

$$\sigma^{\alpha}(x^{\lambda}, -1) = \overset{(0)}{S^{\alpha}} + h\,\overset{(0)}{S^{\beta\alpha}}|_{\beta} - \frac{1}{3}h\,\overset{(1)}{S^{\beta\alpha}}|_{\beta} - hb^{\alpha}_{\beta}\,\overset{(0)}{\tau^{\beta}} + \frac{1}{3}hb^{\alpha}_{\beta}\,\overset{(1)}{\tau^{\beta}} \tag{4.18a}$$

$$\sigma^{\alpha}(x^{\lambda}, +1) = \overset{(0)}{S^{\alpha}} - h\,\overset{(0)}{S^{\beta\alpha}}|_{\beta} - \frac{1}{3}h\,\overset{(1)}{S^{\beta\alpha}}|_{\beta} + hb^{\alpha}_{\beta}\,\overset{(0)}{\tau^{\beta}} + \frac{1}{3}hb^{\alpha}_{\beta}\,\overset{(1)}{\tau^{\beta}} \tag{4.18b}$$

Similarly by introducing formulae (4.14) into (4.17) and making the corresponding cancellations, we obtain

$$\sigma(x^{\lambda}, -1) = \overset{(0)}{\sigma} + h\,\overset{(0)}{\tau^{\alpha}}|_{\alpha} - \frac{1}{3}h\,\overset{(1)}{\tau^{\alpha}}|_{\alpha} + hb^{\alpha}_{\beta}\,\overset{(0)}{S^{\beta}_{\alpha}} - \frac{1}{3}hb^{\alpha}_{\beta}\,\overset{(1)}{S^{\beta}_{\alpha}} \tag{4.19a}$$

$$\sigma(x^{\lambda}, +1) = \overset{(0)}{\sigma} + h\,\overset{(0)}{\tau^{\alpha}}|_{\alpha} - \frac{1}{3}h\,\overset{(1)}{\tau^{\alpha}}|_{\alpha} - hb^{\alpha}_{\beta}\,\overset{(0)}{S^{\beta}_{\alpha}} - \frac{1}{3}hb^{\alpha}_{\beta}\,\overset{(1)}{S^{\beta}_{\alpha}} \tag{4.19b}$$

By means of the above relations, we may now write the face conditions in terms of the coefficients appearing in the expansions (4.7) and (4.13). The face conditions (3.14a) and (3.15a) become

$$\overset{(0)}{S^{\alpha}} + h\,\overset{(0)}{S^{\beta\alpha}}|_{\beta} - \frac{1}{3}h\,\overset{(1)}{S^{\beta\alpha}}|_{\beta} - hb^{\alpha}_{\beta}\,\overset{(0)}{\tau^{\beta}} + \frac{1}{3}hb^{\alpha}_{\beta}\,\overset{(1)}{\tau^{\beta}} = 0 \tag{4.20a}$$

and

$$\overset{(0)}{S}{}^{\alpha} - h\,\overset{(0)}{S}{}^{\beta\alpha}|_{\beta} - \frac{1}{3}h\,\overset{(1)}{S}{}^{\beta\alpha}|_{\beta} + hb^{\alpha}_{\beta}\,\overset{(0)}{\tau}{}^{\beta} + \frac{1}{3}hb^{\alpha}_{\beta}\,\overset{(1)}{\tau}{}^{\beta} = 0 \qquad \textbf{(4.20b)}$$

respectively, while conditions (3.14b) and (3.15b) take the respective forms

$$\overset{(0)}{\sigma} + h\,\overset{(0)}{\tau}{}^{\alpha}|_{\alpha} - \frac{1}{3}h\,\overset{(1)}{\tau}{}^{\alpha}|_{\alpha} + hb^{\alpha}_{\beta}\,\overset{(0)}{S}{}^{\beta}_{\alpha} - \frac{1}{3}hb^{\alpha}_{\beta}\,\overset{(1)}{S}{}^{\beta}_{\alpha} = p^{-} \qquad \textbf{(4.21a)}$$

and

$$\overset{(0)}{\sigma} - h\,\overset{(0)}{\tau}{}^{\alpha}|_{\alpha} - \frac{1}{3}h\,\overset{(1)}{\tau}{}^{\alpha}|_{\alpha} - hb^{\alpha}_{\beta}\,\overset{(0)}{S}{}^{\beta}_{\alpha} - \frac{1}{3}hb^{\alpha}_{\beta}\,\overset{(1)}{S}{}^{\beta}_{\alpha} = p^{+} \qquad \textbf{(4.21b)}$$

The subtraction and addition of relations (4.20) give respectively

$$\overset{(0)}{S}{}^{\beta\alpha}|_{\beta} - b^{\alpha}_{\beta}\,\overset{(0)}{\tau}{}^{\beta} = 0 \qquad \textbf{(4.22)}$$

and

$$\overset{(0)}{S}{}^{\alpha} = \frac{1}{3}h\,\overset{(1)}{S}{}^{\beta\alpha}|_{\beta} - \frac{1}{3}hb^{\alpha}_{\beta}\,\overset{(1)}{\tau}{}^{\beta} \qquad \textbf{(4.23)}$$

while the corresponding combinations of relations (4.21) give

$$2h\,\overset{(0)}{\tau}{}^{\alpha}|_{\alpha} + 2hb^{\alpha}_{\beta}\,\overset{(0)}{S}{}^{\beta}_{\alpha} + p = 0 \qquad \textbf{(4.24)}$$

and

$$\overset{(0)}{\sigma} = \frac{1}{3}h\,\overset{(1)}{\tau}{}^{\alpha}|_{\alpha} + \frac{1}{3}hb^{\alpha}_{\beta}\,\overset{(1)}{S}{}^{\beta}_{\alpha} + \frac{1}{2}p^{*} \qquad \textbf{(4.25)}$$

respectively, where we have used the notation

$$p^{*} = p^{+} + p^{-}, \quad p = p^{+} - p^{-} \qquad \textbf{(4.26)}$$

for the pinching and bending effects of the surface pressures.

Relations (4.23) and 4.25) respectively yield the relations for the completion of the systems (4.10) and (4.14) by providing the omitted form for the case $n = 0$ in each set. Furthermore, the identification of formula (4.23) for $\overset{(0)}{S}{}^{\alpha}$ with the alternative expression (4.8a) leads to the equation relating $\overset{(0)}{\tau}{}^{\alpha}$ to the $\overset{(1)}{S}{}^{\alpha\beta}$ in the form

$$\overset{(0)}{\tau}{}^{\alpha} = \frac{1}{3}h\,\overset{(1)}{S}{}^{\beta\alpha}|_{\beta} \qquad \textbf{(4.27)}$$

thus furnishing the excluded formula for the completion of the system (4.11).

The system of equations (4.22), (4.24) and (4.27) are in fact the well-known integrated equations of shell theory. We may use (4.27) to eliminate $\overset{(0)}{\tau}{}^{\alpha}$ from (4.22) and (4.24) to obtain, respectively

$$\overset{(0)}{S}{}^{\beta\alpha}|_{\beta} - \frac{1}{3} h b^{\alpha}_{\lambda} \overset{(1)}{S}{}^{\beta\lambda}|_{\beta} = 0 \tag{4.28a}$$

and

$$\frac{2}{3} h^2 \overset{(1)}{S}{}^{\alpha\beta}|_{\alpha\beta} + 2hb^{\alpha}_{\beta} \overset{(0)}{S}{}^{\beta}_{\alpha} + p = 0 \tag{4.28b}$$

The alternative system (4.28) constitutes the coupled set of equations to be satisfied by the zero-th and first coefficients in the expansion of the laminar stress tensor.

We may put these equations in more familiar form by introducing notation for the integrated quantities. Defining the stress resultants $N^{\alpha\beta}$ and Q^{α} and the stress couple $M^{\alpha\beta}$ by

$$N^{\alpha\beta} = \int_{-h}^{h} \tau^{\alpha\beta}\, dz = h \int_{-1}^{1} \sigma^{\alpha\beta}\, dt = 2h\, \overset{(1)}{S}{}^{\alpha\beta} \tag{4.29a}$$

$$M^{\alpha\beta} = \int_{-h}^{h} z\, \tau^{\alpha\beta}\, dz = h^2 \int_{-1}^{1} t\, \sigma^{\alpha\beta}\, dt = \frac{2}{3} h^2\, \overset{(1)}{S}{}^{\alpha\beta} \tag{4.29b}$$

$$Q^{\alpha} = \int_{-h}^{h} \tau^{\alpha 3}\, dz = h \int_{-1}^{1} \tau^{\alpha}\, dt = 2h\, \overset{(0)}{\tau}{}^{\alpha} \tag{4.29c}$$

then equations (4.22), (4.24) and (4.27) respectively become

$$N^{\beta\alpha}|_{\beta} - b^{\alpha}_{\beta} Q^{\beta} = 0 \tag{4.30a}$$

$$Q^{\alpha}|_{\alpha} + b^{\alpha}_{\beta} N^{\beta}_{\alpha} + p = 0 \tag{4.30b}$$

$$Q^{\alpha} - M^{\beta\alpha}|_{\beta} = 0 \tag{4.30c}$$

from which the elimination of Q^{α} yields

$$N^{\beta\alpha}|_{\beta} - b^{\alpha}_{\lambda} M^{\beta\lambda}|_{\beta} = 0 \tag{4.31a}$$

$$M^{\alpha\beta}|_{\alpha\beta} + b^{\alpha}_{\beta} N^{\beta}_{\alpha} + p = 0 \tag{4.31b}$$

equivalent to the set (4.28).

4A. Approximate Form of the Transverse Stress Coefficients

The three systems of relations between the coefficients, namely (4.10), (4.11) and (4.14) may now be inverted so as to express the elements appearing in the representations for the transverse stresses, in terms of the laminar stress coefficients. If we introduce the approximation scheme, this can be carried out in a straightforward manner to any desired degree of accuracy. As we shall have use only for the first approximations of these expressions, we shall include only those first order terms: the exclusion of the higher order effects considerably simplifies the inversion.

Starting with the relations for the transverse shear coefficients, the required form for the zero-th term is already given exactly by (4.27) which, for convenience, we repeat here

$$\overset{(0)}{\tau}{}^{\alpha} = \frac{1}{3} h \overset{(1)}{S}{}^{\beta\alpha}|_{\beta} \tag{4A.1}$$

For the first coefficient, we consider (4.11) with $n = 1$: we substitute the above exact form for $\overset{(0)}{\tau}{}^{\beta}$ but the expression for $\overset{(1)}{\tau}{}^{\beta}$ must be introduced from (4.11) with $n = 2$. However, in this latter substitution, the second bracket, since it would give rise to effects of order θ^2, may be omitted: if we also substitute for $\overset{(0)}{S}{}^{\beta\alpha}|_{\beta}$ from (4.22) and rearrange, we have

$$\overset{(1)}{\tau}{}^{\alpha} = \frac{1}{5} h \overset{(2)}{S}{}^{\beta\alpha}|_{\beta} + h^2 b^{\alpha}_{\beta} \Big[\frac{1}{35} \overset{(3)}{S}{}^{\lambda\beta}|_{\lambda} + \frac{4}{15} \overset{(1)}{S}{}^{\lambda\beta}|_{\lambda}\Big] \tag{4A.2}$$

The formula for $\overset{(2)}{\tau}{}^{\alpha}$ is obtained in a similar manner with the added feature that besides omitting the second bracket when introducing $\overset{(3)}{\tau}{}^{\beta}$ from (4.11), we need only consider the first term in (4A.2) when substituting for $\overset{(1)}{\tau}{}^{\beta}$: after combining like terms, we obtain

$$\overset{(2)}{\tau}{}^{\alpha} = h\Big[\frac{1}{7} \overset{(3)}{S}{}^{\beta\alpha}|_{\beta} - \frac{1}{3} \overset{(1)}{S}{}^{\beta\alpha}|_{\beta}\Big] + h^2 b^{\alpha}_{\beta} \Big[\frac{2}{63} \overset{(4)}{S}{}^{\lambda\beta}|_{\lambda} + \frac{1}{7} \overset{(2)}{S}{}^{\lambda\beta}|_{\lambda}\Big] \tag{4A.3}$$

The same pattern is followed for the higher coefficients for which we need use only relation (4.11). We consider the general form and use both the preceding and succeeding formulae to substitute for $\overset{(n-1)}{\tau}{}^{\beta}$ and $\overset{(n+1)}{\tau}{}^{\beta}$ respectively: omitting terms of over θ^2, we rearrange and find

$$\begin{aligned}\overset{(n)}{\tau}{}^{\alpha} &= h\left[\frac{\overset{(n+1)}{S}{}^{\beta\alpha}|_{\beta}}{2n+3} - \frac{\overset{(n-1)}{S}{}^{\beta\alpha}|_{\beta}}{2n-1}\right] \\ &+ h^2 b^{\alpha}_{\beta}\left[\frac{n\ \overset{(n+2)}{S}{}^{\lambda\beta}|_{\lambda}}{(2n+5)(2n+3)} + \frac{3\ \overset{(n)}{S}{}^{\lambda\beta}|_{\lambda}}{(2n+3)(2n-1)} - \frac{(n+1)\ \overset{(n-2)}{S}{}^{\lambda\beta}|_{\lambda}}{(2n-1)(2n-3)}\right], \quad n \geq 3\end{aligned} \tag{4A.4}$$

Next, we derive the corresponding approximate formulae for the coefficients $\overset{(n)}{S}{}^{\alpha}$. If we substitute for $\overset{(1)}{\tau}{}^{\beta}$ from (4A.2) into (4.23) and neglect effects of order θ^2 we obtain for the zero-th coefficients.

$$\overset{(0)}{S}{}^{\alpha} = \frac{1}{3} h \overset{(1)}{S}{}^{\beta\alpha}|_{\beta} - \frac{1}{15} h^2 b^{\alpha}_{\beta} \overset{(2)}{S}{}^{\lambda\beta}|_{\lambda} \qquad \textbf{(4A.5)}$$

For the first coefficient, we use relation (4.10) with $n = 1$, and note the cancellation resulting from equation (4.22): if we then introduce $\overset{(1)}{\tau}{}^{\beta}$ from (4A.3) and neglect effects of order θ^2, we have

$$\overset{(1)}{S}{}^{\alpha} = \frac{1}{5} h \overset{(2)}{S}{}^{\beta\alpha}|_{\beta} - h^2 b^{\alpha}_{\beta} [\frac{1}{35} \overset{(3)}{S}{}^{\lambda\beta}|_{\lambda} - \frac{1}{15} \overset{(1)}{S}{}^{\lambda\beta}|_{\lambda}] \qquad \textbf{(4A.6)}$$

For $\overset{(2)}{S}{}^{\alpha}$ we take (4.10) with $n = 2$ and introduce $\overset{(1)}{\tau}{}^{\beta}$ from (4A.2) and $\overset{(3)}{\tau}{}^{\beta}$ from (4A.4) with $n = 3$: omitting terms of order θ^2, there follows

$$\overset{(2)}{S}{}^{\alpha} = h[\frac{1}{7} \overset{(3)}{S}{}^{\beta\alpha}|_{\beta} - \frac{1}{3} \overset{(1)}{S}{}^{\beta\alpha}|_{\beta}] - h^2 b^{\alpha}_{\beta} [\frac{1}{63} \overset{(4)}{S}{}^{\lambda\beta}|_{\lambda} - \frac{2}{21} \overset{(2)}{S}{}^{\lambda\beta}|_{\lambda}] \qquad \textbf{(4A.7)}$$

The determination of $\overset{(3)}{S}{}^{\alpha}$ involves the introduction of $\overset{(2)}{\tau}{}^{\beta}$ from (4A.3): however, since the second bracket in (4A.3) would give rise to effects of order θ^2 its special form would not affect the result. Hence, the formula for $\overset{(3)}{S}{}^{\alpha}$ follows the general pattern for the higher coefficients, for which we use (4.10) and introduce $\overset{(n+1)}{\tau}{}^{\beta}$ and $\overset{(n-1)}{\tau}{}^{\beta}$ from (4A.4): after omitting effects of order θ^2, we rearrange and find

$$\overset{(n)}{S}{}^{\alpha} = h \left[\frac{\overset{(n+1)}{S}{}^{\beta\alpha}|_{\beta}}{2n+3} - \frac{\overset{(n-1)}{S}{}^{\beta\alpha}|_{\beta}}{2n-1} \right]$$

$$- h^2 b^{\alpha}_{\beta} \left[\frac{\overset{(n+2)}{S}{}^{\lambda\beta}|_{\lambda}}{(2n+5)(2n+3)} - \frac{2\overset{(n)}{S}{}^{\lambda\beta}|_{\lambda}}{(2n+3)(2n-1)} + \frac{\overset{(n-2)}{S}{}^{\lambda\beta}|_{\lambda}}{(2n-1)(2n-3)} \right], \; n \geq 3 \qquad \textbf{(4A.8)}$$

Finally, we write the expressions for the coefficients appearing in the representations for the transverse normal stress. It is also convenient at this point to relabel the dummy indices appearing in these formulae. If we introduce $\overset{(1)}{\tau}{}^{\alpha}$ from (4A.2) into formula (4.25) for the zero-th coefficient, we have

$$\overset{(0)}{\sigma} = \frac{1}{15} \overset{(2)}{S}{}^{\lambda\mu}|_{\lambda\mu} + \frac{1}{3} h b^{\lambda}_{\mu} \overset{(1)}{S}{}^{\mu}_{\lambda} + \frac{1}{2} p^* + h^3 [\frac{4}{15} (b^{\lambda}_{\rho} \overset{(1)}{S}{}^{\mu\rho}|_{\mu})|_{\lambda} + \frac{1}{105} (b^{\lambda}_{\rho} \overset{(3)}{S}{}^{\mu\rho}|_{\mu})|_{\lambda}] \qquad \textbf{(4A.9)}$$

For the first coefficient, we consider formula (4.14) with $n = 1$; after introducing $\overset{(0)}{\tau}{}^\alpha$ from (4A.1) and $\overset{(2)}{\tau}{}^\alpha$ from (4A.3), we rearrange the terms: if we then use equation (4.28b) to eliminate $\overset{(1)}{S}{}^{\lambda\mu}|_{\lambda\mu}$, there follows

$$\overset{(0)}{\sigma} = \frac{1}{35} h^2 \overset{(3)}{S}{}^{\lambda\mu}|_{\lambda\mu} + \frac{3}{5} p + \frac{1}{5} h b^\lambda_\mu (\overset{(0)}{S}{}^\mu_\lambda + \overset{(2)}{S}{}^\mu_\lambda)$$
$$+ h^3 [\frac{1}{35} (b^\lambda_\rho \overset{(2)}{S}{}^{\mu\rho}|_\mu)|_\lambda + \frac{2}{315} (b^\lambda_\rho \overset{(4)}{S}{}^{\mu\rho}|_\mu)|_\lambda] \qquad \textbf{(4A.10)}$$

In a similar manner, we use formulae (4A.2) and (4A.4) to substitute for $\overset{(1)}{\tau}{}^\alpha$ and $\overset{(3)}{\tau}{}^\alpha$ respectively in relation (4.14) with $n = 2$, which on rearrangement then becomes

$$\overset{(2)}{\sigma} = h^2 [- \frac{2}{21} \overset{(2)}{S}{}^{\lambda\mu}|_{\lambda\mu} + \frac{1}{63} \overset{(4)}{S}{}^{\lambda\mu}|_{\lambda\mu}] - h b^\lambda_\mu [\frac{1}{3} \overset{(1)}{S}{}^\mu_\lambda - \frac{1}{7} \overset{(3)}{S}{}^\mu_\lambda]$$
$$- h^3 [\frac{8}{63} (b^\lambda_\rho \overset{(1)}{S}{}^{\mu\rho}|_\mu)|_\lambda - \frac{1}{231} (b^\lambda_\rho \overset{(5)}{S}{}^{\mu\rho}|_\mu)|_\lambda] \qquad \textbf{(4A.11)}$$

After introducing $\overset{(2)}{\tau}{}^\alpha$ from (4A.3) and $\overset{(4)}{\tau}{}^\alpha$ from (4A.4) into the formula for $\overset{(3)}{\sigma}$ as given, by (4.14), we may again use equation (4.28b) to eliminate $\overset{(1)}{S}{}^{\lambda\mu}|_{\lambda\mu}$: after some rearrangement, we obtain

$$\overset{(3)}{\sigma} = h^2 [- \frac{1}{45} \overset{(3)}{S}{}^{\lambda\mu}|_{\lambda\mu} + \frac{1}{99} \overset{(5)}{S}{}^{\lambda\mu}|_{\lambda\mu}] - \frac{1}{10} p - h b^\lambda_\mu [\frac{1}{5} \overset{(0)}{S}{}^\mu_\lambda + \frac{1}{5} \overset{(2)}{S}{}^\mu_\lambda - \frac{1}{9} \overset{(4)}{S}{}^\mu_\lambda]$$
$$- h^3 [\frac{2}{45} (b^\lambda_\rho \overset{(2)}{S}{}^{\mu\rho}|_\mu)|_\lambda + \frac{1}{495} (b^\lambda_\rho \overset{(4)}{S}{}^{\mu\rho}|_\mu)|_\lambda - \frac{1}{1287} (b^\lambda_\rho \overset{(6)}{S}{}^{\mu\rho}|_\mu)|_\lambda] \qquad \textbf{(4A.12)}$$

For the higher coefficients, it suffices to substitute for $\overset{(n+1)}{\tau}{}^\alpha$ and $\overset{(n-1)}{\tau}{}^\alpha$ from (4A.4) into (4.14) and rearrange: we find

$$\overset{(n)}{\sigma} = h^2 \left[\frac{\overset{(n-2)}{S}{}^{\lambda\mu}|_{\lambda\mu}}{(2n-3)(2n-1)} - \frac{2 \overset{(n)}{S}{}^{\lambda\mu}|_{\lambda\mu}}{(2n-1)(2n+3)} + \frac{\overset{(n+2)}{S}{}^{\lambda\mu}|_{\lambda\mu}}{(2n+3)(2n+5)} \right]$$
$$- h b^\lambda_\mu \left[\frac{\overset{(n-1)}{S}{}^\mu_\lambda}{2n-1} - \frac{\overset{(n+1)}{S}{}^\mu_\lambda}{2n+3} \right]$$
$$- h^3 \left[\frac{n (b^\lambda_\rho \overset{(n-3)}{S}{}^{\mu\rho}|_\mu)|_\lambda}{(2n-5)(2n-3)(2n-1)} - \frac{(n+3)(b^\lambda_\rho \overset{(n-1)}{S}{}^{\mu\rho}|_\mu)|_\lambda}{(2n-3)(2n-1)(2n+3)} \right.$$

$$+\frac{(n-2)(b^{\lambda}_{\rho}\overset{(n+1)}{S^{\mu\rho}}|_{\mu})|_{\lambda}}{(2n-1)(2n+3)(2n+5)}-\frac{(n+1)(b^{\lambda}_{\rho}\overset{(n+3)}{S^{\mu\rho}}|_{\mu})|_{\lambda}}{(2n+3)(2n+5)(2n+7)}\Bigg],\quad n\geq 4 \qquad \textbf{(4A.13)}$$

thus completing the determination of the transverse elements in terms of the laminar coefficients.

5. The Edge Boundary Conditions

Consistent with the representations derived for the stress components, it is appropriate to reformulate the edge conditions on the stresses so as to express equivalent requirements on the elements appearing in the Legendre expansions. Conditions (3.16a) are then transformed into a sequence of constraints on the laminar stress coefficients $\overset{(n)}{S^{\alpha\beta}}$ while conditions (3.16b) become a sequence of requirements on shear stress coefficients $\overset{(n)}{\tau^{\alpha}}$.

The edge conditions thus obtained for the coefficients $\overset{(0)}{S^{\alpha\beta}}$, $\overset{(1)}{S^{\alpha\beta}}$ and $\overset{(0)}{\tau^{\alpha}}$ are associated with the equilibrium equations (4.22), (4.24) and (4.27). By means of (4.27), we can transform the condition on $\overset{(0)}{\tau^{\alpha}}$ into a further equivalent condition on $\overset{(1)}{S^{\alpha\beta}}$ and thus complete the set of conditions appropriate for the alternate system of equilibrium equations (4.28).

The derived conditions on the remaining coefficients $\{\overset{(n)}{S^{\alpha\beta}},\ n\geq 2:\ \overset{(n)}{\tau^{\alpha}},\ n\geq 1\}$ complement the system of equations for the higher coefficients to be described later. As these latter equations will be formulated exclusively in terms of the laminar coefficients, it is desirable that the associated edge conditions also be so defined: this will be done by transforming the conditions on the $\overset{(n)}{\tau^{\alpha}}$ into equivalent conditions on the $\overset{(n)}{S^{\alpha\beta}}$.

If we write the edge values of the coefficients

$$\overline{\overset{(n)}{S^{\alpha\beta}}}(\eta)=\left[\overset{(n)}{S^{\alpha\beta}}(x^{\lambda})\right]_{\overline{x^{\lambda}=x^{\lambda}(\eta)}},\quad \overline{\overset{(n)}{\tau^{\alpha}}}(\eta)=\left[\overset{(n)}{\tau^{\alpha}}(x^{\lambda})\right]_{\overline{x^{\lambda}=x^{\lambda}(\eta)}} \qquad \textbf{(5.1a,b)}$$

then for the edge values (3.8) of the stress components, we have the representations

$$\overline{\sigma^{\alpha\beta}}(\eta,t)=\sum_{n=0}^{\infty}\overline{\overset{(n)}{S^{\alpha\beta}}}(\eta)\,P_n(t),\quad \overline{\tau^{\alpha}}(\eta,t)=\sum_{n=0}^{\infty}\overline{\overset{(n)}{\tau^{\alpha}}}(\eta)\,P_n(t) \qquad \textbf{(5.2a,b)}$$

Expanding the components of the applied stress distribution vector in the form

$$T^{\alpha}(\eta,t)=\sum_{n=0}^{\infty}\overset{(n)}{T^{\alpha}}(\eta)\,P_n(t),\quad T(\eta,t)=\sum_{n=0}^{\infty}\overset{(n)}{T}(\eta)\,P_n(t) \qquad \textbf{(5.3a,b)}$$

it follows directly that the three edge constraints (3.16) are equivalent to the three sequences of conditions on the coefficients

$$x^{\lambda} = x^{\lambda}(\eta): \begin{cases} \overline{\overset{(n)}{S}{}^{\beta\alpha}(\eta)\, n_{\beta}(\eta)} = \overset{(n)}{T}{}^{\alpha}(\eta), n \geq 0 & \textbf{(5.4a)} \\ \overline{\overset{(n)}{\tau}{}^{\alpha}(\eta)\, n_{\alpha}(\eta)} = \overset{(n)}{T}(\eta), n \geq 0 & \textbf{(5.4b)} \end{cases}$$

In particular we have the conditions

$$\overline{\overset{(0)}{S}{}^{\beta\alpha}(\eta)\, n_{\beta}(\eta)} = \overset{(0)}{T}{}^{\alpha}(\eta), \quad \overline{\overset{(1)}{S}{}^{\beta\alpha}(\eta)\, n_{\beta}(\eta)} = \overset{(1)}{T}{}^{\alpha}(\eta), \quad \overline{\overset{(0)}{\tau}{}^{\alpha}(\eta)\, n_{\alpha}(\eta)} = \overset{(0)}{T}(\eta)$$

(5.5a,b,c)

associated with the system of equilibrium equations (4.22), (4.24) and (4.27).

In transforming conditions (5.4b) on the coefficients $\overset{(n)}{\tau}{}^{\alpha}$ into an equivalent set of conditions on the $\overset{(n)}{S}{}^{\alpha\beta}$, we use relations (4A.1) to (4A.4). If we write successively the formulae for the transverse shear stress coefficients with even index n, then from (4A.1), (4A.3) and (4A.4) we have the following sequence of relations for the $\overset{(2k)}{\tau}{}^{\alpha}$

$$\overset{(0)}{\tau}{}^{\alpha} = \frac{1}{3} h \overset{(1)}{S}{}^{\beta\alpha}|_{\beta}$$

$$\overset{(2)}{\tau}{}^{\alpha} = h[\tfrac{1}{7} \overset{(3)}{S}{}^{\beta\alpha}|_{\beta} - \tfrac{1}{3} \overset{(1)}{S}{}^{\beta\alpha}|_{\beta}] + h^2 b^{\alpha}_{\beta}[\tfrac{2}{63} \overset{(4)}{S}{}^{\lambda\beta}|_{\lambda} + \tfrac{1}{7} \overset{(2)}{S}{}^{\lambda\beta}|_{\lambda}]$$

$$\overset{(4)}{\tau}{}^{\alpha} = h[\tfrac{1}{11} \overset{(5)}{S}{}^{\beta\alpha}|_{\beta} - \tfrac{1}{7} \overset{(3)}{S}{}^{\beta\alpha}|_{\beta}] + h^2 b^{\alpha}_{\beta}[\tfrac{4}{143} \overset{(6)}{S}{}^{\lambda\beta}|_{\lambda} + \tfrac{3}{77} \overset{(4)}{S}{}^{\lambda\beta}|_{\lambda} - \tfrac{1}{7} \overset{(2)}{S}{}^{\lambda\beta}|_{\lambda}]$$

$$\vdots$$

$$\overset{(2k)}{\tau}{}^{\alpha} = h\left[\frac{\overset{(2k+1)}{S}{}^{\beta\alpha}|_{\beta}}{4k+3} - \frac{\overset{(2k-1)}{S}{}^{\beta\alpha}|_{\beta}}{4k-1}\right]$$

$$+ h^2 b^{\alpha}_{\beta}\left[\frac{2k \ \overset{(2k+2)}{S}{}^{\lambda\beta}|_{\lambda}}{(4k+5)(4k+3)} + \frac{3 \ \overset{(2k)}{S}{}^{\lambda\beta}|_{\lambda}}{(4k+3)(4k-1)} - \frac{(2k+1) \ \overset{(2k-2)}{S}{}^{\lambda\beta}|_{\lambda}}{(4k-1)(4k-3)}\right]$$

which, when added, give, for $k \geq 1$,

$$\sum_{i=0}^{k} \overset{(2i)}{\tau}{}^{\alpha} = \frac{1}{4k+3}\, h \overset{(2k+1)}{S}{}^{\beta\alpha}|_{\beta} + \frac{2k}{(4k+5)(4k+3)}\, h^2 b^{\alpha}_{\beta} \overset{(2k+3)}{S}{}^{\lambda\beta}|_{\lambda}$$
$$+ \frac{2k+3}{(4k+3)(4k+1)}\, h^2 b^{\alpha}_{\beta} \overset{(2k)}{S}{}^{\lambda\beta}|_{\lambda}, \quad k \geq 1 \qquad (5.6)$$

while the case $k = 0$ is covered by (4A.1): actually (5.6) can be used for this case also if we note that the second term vanishes and that the third is of order θ^2 by virtue of (4.22). By multiplying by the factor $(4k+3)$, we may write these relations in the alternative form

$$h \overset{(1)}{S}{}^{\beta\alpha}|_{\beta} = 3 \overset{(0)}{\tau}{}^{\alpha} \qquad (5.7a)$$

$$h \overset{(2k+1)}{S}{}^{\beta\alpha}|_{\beta} + \frac{2k}{4k+5}\, h^2 b^{\alpha}_{\beta} \overset{(2k+2)}{S}{}^{\lambda\beta}|_{\lambda} + \frac{2k+3}{4k+1}\, h^2 b^{\alpha}_{\beta} \overset{(2k)}{S}{}^{\lambda\beta}|_{\lambda} = (4k+3) \sum_{i=0}^{k} \overset{(2i)}{\tau}{}^{\alpha}, \quad k \geq 1 \qquad (5.7b)$$

Similarly, using relations (4A.2) and (4A.4) we may list successively the formulae for the transverse shear coefficients of odd index n: performing a corresponding summation on these relations for the $\overset{(2k+1)}{\tau}{}_{\alpha}$, we obtain for $k \geq 1$,

$$\sum_{i=0}^{(k-1)} \overset{(2i+1)}{\tau}{}^{\alpha} = \frac{1}{4k+1}\, h \overset{(2k)}{S}{}^{\beta\alpha}|_{\beta} + \frac{2k-1}{(4k+3)(4k+1)}\, h^2 b^{\alpha}_{\beta} \overset{(2k+1)}{S}{}^{\lambda\beta}|_{\lambda}$$
$$+ \frac{2k+2}{(4k+1)(4k-1)}\, h^2 b^{\alpha}_{\beta} \overset{(2k-1)}{S}{}^{\lambda\beta}|_{\lambda} \quad k \geq 1, \qquad (5.8)$$

which, on multiplication by the factor $(4k+1)$, assumes the alternate form.

$$h \overset{(2k)}{S}{}^{\beta\alpha}|_{\beta} + \frac{2k-1}{4k+3}\, h^2 b^{\alpha}_{\beta} \overset{(2k+1)}{S}{}^{\lambda\beta}|_{\lambda} + \frac{2k+2}{4k-1}\, h^2 b^{\alpha}_{\beta} \overset{(2k-1)}{S}{}^{\lambda\beta}|_{\lambda} = (4k+1) \sum_{i=0}^{k-1} \overset{(2i+1)}{\tau}{}^{\alpha}, \quad k \geq 1, \qquad (5.9)$$

Relations (5.7b) and (5.9) are the analogs, for the higher coefficients of equation (5.7a).

We shall use relations (5.7) and (5.9) only at thin edge values; if we set

$$\overset{(2k+1)}{D}(\eta) = (4k+3)\sum_{i=0}^{(k)} \overset{(2i)}{T}(\eta),\ k \geq 0 \tag{5.10a}$$

$$\overset{(2k)}{D}(\eta) = (4k+1)\sum_{i=0}^{(k-1)} \overset{(2i+1)}{T}(\eta),\ k \geq 1 \tag{5.10b}$$

we see that (5.4) is equivalent to two sets of conditions, the first of which are

$$\overline{\overset{(0)}{S}{}^{\beta\alpha}}\, n_\beta = \overset{(0)}{T}{}^{\alpha},\quad \overline{\overset{(1)}{S}{}^{\beta\alpha}}\, n_\beta = \overset{(1)}{T}{}^{\alpha},\quad h\,\overline{\overset{(1)}{S}{}^{\beta\alpha}|_\beta}\, n_\alpha = \overset{(1)}{D} \tag{5.11a,b,c}$$

while the second set consists of the remaining condition

$$\overline{\overset{(n)}{S}{}^{\beta\alpha}}\, n_\beta = \overset{(n)}{T}{}^{\alpha},\ n \geq 2 \tag{5.12a}$$

$$h\,\overline{\overset{(n)}{S}{}^{\beta\alpha}|_\beta}\, n_\alpha + \frac{n-1}{2n+3} h^2 \overline{b^\alpha_\beta}\; \overline{\overset{(n+1)}{S}{}^{\lambda\beta}|_\lambda}\, n_\alpha + \frac{n+2}{2n-1} h^2 \overline{b^\alpha_\beta}\; \overline{\overset{(n-1)}{S}{}^{\lambda\beta}|_\lambda}\, n_\alpha = \overset{(n)}{D},\ n \geq 2 \tag{5.12b}$$

in which the bar is also used to designate the edge values of the components of the second fundamental form.

The set (5.11) equivalent to conditions (5.5) are associated with the alternate form (4.28) of the equilibrium equation. The second set (5.12) complements the system of differential equations for the higher laminar stress coefficients $\{\overset{(n)}{S}{}^{\alpha\beta}, n \geq 2\}$ to be formulated later. However, in the first condition of the sequence (5.12b) there also occurs the laminar coefficient $\overset{(1)}{S}{}^{\alpha\beta}$: thus, it would be more appropriate to write (5.12b) in the form

$$\left.\begin{aligned} &h\,\overline{\overset{(2)}{S}{}^{\beta\alpha}|_\beta}\, n_\alpha + \tfrac{1}{5} h^2 \overline{b^\alpha_\beta}\; \overline{\overset{(3)}{S}{}^{\lambda\beta}|_\lambda}\, n_\alpha = \overset{(2)}{D} - \tfrac{4}{3} h^2 \overline{b^\alpha_\beta}\; \overline{\overset{(1)}{S}{}^{\lambda\beta}|_\lambda} \\ &h\,\overline{\overset{(n)}{S}{}^{\beta\alpha}|_\beta}\, n_\alpha + \frac{n-1}{2n+3} h^2 \overline{b^\alpha_\beta}\; \overline{\overset{(n+1)}{S}{}^{\lambda\beta}|_\lambda}\, n_\alpha + \frac{n+2}{2n-1} h^2 \overline{b^\alpha_\beta}\; \overline{\overset{(n-1)}{S}{}^{\lambda\beta}|_\lambda} = \overset{(n)}{D},\ n \geq 3 \end{aligned}\right\} \tag{5.12*}$$

in which quantity $\overset{(1)}{S}{}^{\alpha\beta}$ is now presumed determined. Thus, in the case of the shell, the principal problem must be solved prior to the analysis of the residual problem: this is further discussed later.

6. Representations for the Displacement Components

The Legendre series for the displacement quantities will follow from the integration of the constitutive relations (3.12a,b). In this step of the procedure, we shall again use formula (4.3) and, due to the terms arising directly from the curvature, we shall also need the integrated form of relation (4.1). Using formula (4.2) to express the right hand side of (4.1) in terms of derivatives of Legendre functions, we find

$$tP_n(t) = \frac{(n+1)\dot{P}_{n+2}(t)}{(2n+3)(2n+1)} + \frac{\dot{P}_n(t)}{(2n+3)(2n-1)} - \frac{n\dot{P}_{n-2}(t)}{(2n+1)(2n-1)} \tag{6.1}$$

which, apart from an additive term independent of t, yields

$$\int tP_n(t) = \frac{(n+1)P_{n+2}(t)}{(2n+3)(2n+1)} + \frac{P_n(t)}{(2n+3)(2n-1)} - \frac{nP_{n-2}(t)}{(2n+1)(2n-1)} \tag{6.2}$$

We introduce the dimensionless constants Λ_* and Λ by setting

$$\Lambda_* = \frac{E}{E_z}, \quad \Lambda(1+\nu) = \frac{E}{B} \tag{6.3a,b}$$

so that, with a little rearrangement, the constitutive relations (3.12a,b) may be written

$$w,_t = \frac{h}{E}[(1 + 2hHt)(\Lambda_*\sigma - \nu_*\sigma^\lambda_\lambda) + \nu_* thb^\lambda_\mu\sigma^\mu_\lambda] \tag{6.4a}$$

$$u_{\alpha,t} = 2\frac{h}{E}\Lambda(1+\nu)[(1 + 2hHt)\sigma_\alpha - thb^\lambda_\alpha\sigma_\lambda] - hw,_\alpha - 2hb^\lambda_\alpha u_\lambda \tag{6.4b}$$

where the purpose of the transpositions made in the latter equation will be recognized presently. Moreover, by substituting for $u_{\alpha,t}$ from (6.4b) into formula (3.12b), we obtain, for the transverse components of rotation, the alternate expression

$$\omega_\alpha = \frac{1}{E}\Lambda(1+\nu)[(1 + 2hHt)\sigma_\alpha - thb^\lambda_\alpha\sigma_\lambda] - w,_\alpha - b^\lambda_\alpha u_\lambda \tag{6.5a}$$

while we retain the original form

$$\psi = \frac{1}{2}(u_{1,2} - u_{2,1}) \tag{6.5b}$$

for the laminar rotation.

If we insert the expansion (4.4) for $\sigma^{\alpha\beta}$, together with the related representation (4.13) for σ, into the right hand side, we obtain the series form of the differential equation (6.4a), namely

$$w,_t = \frac{h}{E}[\sum_{n=0}^{\infty}(\Lambda_* \overset{(n)}{\sigma} - \nu_* \overset{(n)}{S^\lambda_\lambda})P_n + 2hH\sum_{n=0}^{\infty}(\Lambda_* \overset{(n)}{\sigma} - \nu_* \overset{(n)}{S^\lambda_\lambda})tP_n + \nu_* hb^\lambda_\mu \sum_{n=0}^{\infty} \overset{(n)}{S^\mu_\lambda} tP_n] \tag{6.6}$$

The application of the integration formulae (4.3) and (6.5) yields the Legendre series for the transverse displacement in the form

$$w = \sum_{n=0}^{\infty} \overset{(n)}{w}(x^\lambda)\, P_n(t) \tag{6.7}$$

in which $\overset{(n)}{w}$ is an unknown function introduced by the integration and the higher coefficients are given by

$$\overset{(n)}{w} = -\frac{h}{E}\left\{ \frac{\Lambda_* \overset{(n+1)}{\sigma} - \nu_* \overset{(n+1)}{S_\lambda^\lambda}}{2n+3} - \frac{\Lambda_* \overset{(n-1)}{\sigma} - \nu_* \overset{(n-1)}{S_\lambda^\lambda}}{2n-1} \right.$$

$$+ 2hH \left[\frac{(n+2)(\Lambda_* \overset{(n+2)}{\sigma} - \nu_* \overset{(n+2)}{S_\lambda^\lambda})}{(2n+5)(2n+3)} - \frac{\Lambda_* \overset{(n)}{\sigma} - \nu_* \overset{(n)}{S_\lambda^\lambda}}{(2n+3)(2n-1)} - \frac{(n-1)(\Lambda_* \overset{(n-2)}{\sigma} - \nu_* \overset{(n-2)}{S_\lambda^\lambda})}{(2n-1)(2n-3)} \right]$$

$$\left. + \nu_* h b_\mu^\lambda \left[\frac{(n+2)\, \overset{(n+2)}{S_\lambda^\mu}}{(2n+5)(2n+3)} - \frac{\overset{(n)}{S_\lambda^\mu}}{(2n+3)(2n-1)} - \frac{(n-1)\, \overset{(n-2)}{S_\lambda^\mu}}{(2n-1)(2n-3)} \right] \right\}$$

$$n \geq 1 \tag{6.8}$$

Although the last terms in each of the square brackets have index $(n-2)$, their interpretation for $n = 1$ does not require any special attention since, in each case, they are multiplied by the factor $(n-1)$.

The most convenient way to effect the integration of the differential equation for u_α is to anticipate the Legendre series for the laminar displacements, namely,

$$u_\alpha = \sum_{n=0}^{\infty} \overset{(n)}{u_\alpha}(x^\lambda)\, P_n(t) \tag{6.9}$$

in the transposed terms involving the u_α on the extreme right of (6.4b). The integration then leads to a sequence of equations for the determination of the coefficients, which can be readily solved to the desired degree of accuracy.

If we introduce the expansion (4.7b) for σ_α, together with the representations (6.7) and (6.9) for the displacement quantities, into the right hand side of (6.4b), we obtain the series form of the differential equations for the laminar displacements, namely

$$u_{\alpha,t} = 2\frac{h}{E}\Lambda(1+\nu)[\sum_{n=0}^{\infty} \overset{(n)}{S}_{\alpha} P_n + 2hH \sum_{n=0}^{\infty} \overset{(n)}{S}_{\alpha} tP_n - hb_{\alpha}^{\lambda} \sum_{n=0}^{\infty} \overset{(n)}{S}_{\lambda} tP_n]$$

$$- h \sum_{n=0}^{\infty} \overset{(n)}{w}|_{\alpha} P_n - 2hb_{\alpha}^{\lambda} \sum_{n=0}^{\infty} \overset{(n)}{u}_{\lambda} P_n \quad \textbf{(6.10)}$$

where for the partial derivatives of the transverse displacement coefficients we have used the symbol for covariant rather than partial differentiation: this is permissible since, for invariant quantities, these operations are indistinguishable. By the application of formulae (4.3) and (6.2), we may now integrate equation (6.10) and, after some rearrangement, we obtain, for the laminar displacements, Legendre representations of the form (6.9) in which the zero-th coefficient $\overset{(0)}{u}_{\alpha}$ remains unspecified and the higher coefficients are to be determined from the sequence of equations

$$\overset{(n)}{u}_{\alpha} = -2\frac{h}{E}\Lambda(1+\nu)\left[\frac{\overset{(n+1)}{S}_{\alpha}}{2n+3} - \frac{\overset{(n-1)}{S}_{\alpha}}{2n-1}\right] + \left[\frac{h\ \overset{(n+1)}{w}|_{\alpha}}{2n+3} - \frac{h\ \overset{(n-1)}{w}|_{\alpha}}{2n-1}\right]$$

$$+ 2hb_{\alpha}^{\lambda}\left[\frac{\overset{(n+1)}{u}_{\lambda}}{2n+3} - \frac{\overset{(n-1)}{u}_{\lambda}}{2n-1}\right]$$

$$-2\frac{h}{E}\Lambda(1+\nu)\left\{2hH\left[\frac{(n+2)\ \overset{(n+2)}{S}_{\alpha}}{(2n+5)(2n+3)} - \frac{\overset{(n)}{S}_{\alpha}}{(2n+3)(2n-1)} - \frac{(n-1)\ \overset{(n-2)}{S}_{\alpha}}{(2n-1)(2n-3)}\right]\right.$$

$$\left. - hb_{\alpha}^{\lambda}\left[\frac{(n+2)\ \overset{(n+2)}{S}_{\lambda}}{(2n+5)(2n+3)} - \frac{\overset{(n)}{S}_{\lambda}}{(2n+3)(2n-1)} - \frac{(n-1)\ \overset{(n-2)}{S}_{\lambda}}{(2n-1)(2n-3)}\right]\right\}$$

$$, n \geq 1 \quad \textbf{(6.11)}$$

If, into each of equations (6.11), we introduce $\overset{(n-1)}{u}_{\lambda}$ and $\overset{(n+1)}{u}_{\lambda}$ from the preceding and the succeeding equations, respectively, then on neglecting effects of order θ^2, we obtain the explicit expressions for the laminar displacement coefficients in the form

$$\overset{(1)}{u}_\alpha = -h\,\overset{(0)}{w}|_\alpha - 2hb^\lambda_\alpha \overset{(0)}{u}_\lambda + 2\frac{h}{E}\Lambda(1+\nu)[\,\overset{(0)}{S}_\alpha - \frac{1}{5}\overset{(2)}{S}_\alpha]$$

$$+\,2\frac{h}{E}\Lambda(1+\nu)\left[2hH(\frac{1}{15}\overset{(1)}{S}_\alpha - \frac{1}{35}\overset{(3)}{S}_\alpha) - hb^\lambda_\alpha(\frac{1}{15}\overset{(1)}{S}_\lambda - \frac{1}{35}\overset{(3)}{S}_\lambda)\right]$$

$$+\,\frac{1}{5}h\left[\overset{(2)}{w}|_\alpha - 2hb^\lambda_\alpha(\frac{1}{3}\overset{(1)}{w} - \frac{1}{7}\overset{(3)}{w})|_\lambda\right] \qquad \textbf{(6.12a)}$$

$$\overset{(n)}{u}_\alpha = -2\frac{h}{E}\Lambda(1+\nu)\left[\frac{\overset{(n+1)}{S}_\alpha}{2n+3} - \frac{\overset{(n-1)}{S}_\alpha}{2n-1}\right] + \left[\frac{h\,\overset{(n+1)}{w}|_\alpha}{2n+3} - \frac{h\,\overset{(n-1)}{w}|_\alpha}{2n-1}\right]$$

$$+\,2hb^\lambda_\alpha\left[\frac{h\,\overset{(n+2)}{w}|_\lambda}{(2n+5)(2n-3)} - \frac{2h\,\overset{(n)}{w}|_\lambda}{(2n+3)(2n-1)} + \frac{\overset{(n-2)}{w}|_\lambda}{(2n-1)(2n-3)}\right]$$

$$-\,2\frac{h}{E}\Lambda(1+\nu)\left\{2hH\left[\frac{(n+2)\,\overset{(n+2)}{S}_\alpha}{(2n+5)(2n+3)} - \frac{\overset{(n)}{S}_\alpha}{(2n+3)(2n-1)} - \frac{(n-1)\,\overset{(n-2)}{S}_\alpha}{(2n-1)(2n-3)}\right]\right.$$

$$\left.-\,hb^\lambda_\alpha\left[\frac{n\,\overset{(n+2)}{S}_\lambda}{(2n+5)(2n+3)} + \frac{3\,\overset{(n)}{S}_\lambda}{(2n+3)(2n-1)} - \frac{(n+1)\,\overset{(n-2)}{S}_\lambda}{(2n-1)(2n-3)}\right]\right\}$$

$$n \ge 2 \qquad \textbf{(6.12b)}$$

By substituting into expressions (6.12), the formulae (4A.5) to (4A.8) for the $\overset{(n)}{S}_\alpha$, as well as expressions (6.8) for the $\overset{(n)}{w}$, together with the implied formulae (4A.9) to (4A.13) for the $\overset{(n)}{\sigma}$, we could obtain the set of formulae expressing the laminar displacement coefficients directly in terms of the laminar stress coefficients.

From the Legendre series for the stress and displacement quantities, we now derive the corresponding representations for the rotation components (6.5). Utilizing the expansion (4.7b) for σ_α, together with the series (6.7) and (6.9) for the displacement quantities, formula (6.5a) becomes

$$\omega_\alpha = \frac{1}{E}\Lambda(1+\nu)\,[\sum_{n=0}^{\infty} \overset{(n)}{S}_\alpha P_n + 2hH \sum_{n=0}^{\infty} \overset{(n)}{S}_\alpha tP_n - hb_\alpha^\lambda \sum_{n=0}^{\infty} \overset{(n)}{S}_\lambda tP_n]$$

$$- \sum_{n=0}^{\infty} \overset{(n)}{w}|_\alpha P_n - b_\alpha^\lambda \sum_{n=0}^{\infty} \overset{(n)}{u}_\lambda P_n \quad \textbf{(6.13)}$$

where again, we have used the covariant sign in the differentiation of the transverse displacement coefficients. Applying the recursion formula (4.1) and rearranging, we obtain the Legendre series for the transverse rotations

$$\omega_\alpha = \sum \overset{(n)}{\omega}_\alpha(x^\lambda)\, P_n(t) \qquad \textbf{(6.14)}$$

in which the coefficients are given by

$$\overset{(0)}{\omega}_\alpha = \frac{1}{E}\Lambda(1+\nu)\left[\overset{(0)}{S}_\alpha + \frac{2}{3}hH\overset{(1)}{S}_\alpha - \frac{1}{3}hb_\alpha^\lambda \overset{(1)}{S}_\lambda\right] - h\overset{(0)}{w}|_\alpha - b_\alpha^\lambda \overset{(0)}{u}_\lambda \qquad \textbf{(6.15a)}$$

$$\overset{(n)}{\omega}_\alpha = \frac{1}{E}\Lambda(1+\nu)\left[\overset{(n)}{S}_\alpha + 2hH\left(\frac{n}{2n-1}\overset{(n-1)}{S}_\alpha + \frac{n+1}{2n+3}\overset{(n+1)}{S}_\alpha\right)\right.$$

$$\left. - hb_\alpha^\lambda\left(\frac{n}{2n-1}\overset{(n-1)}{S}_\lambda + \frac{n+1}{2n+3}\overset{(n+1)}{S}_\lambda\right)\right]$$

$$- h\overset{(n)}{w}|_\alpha - b_\alpha^\lambda \overset{(n)}{u}|_\lambda\,,\ n \geq 1 \qquad \textbf{(6.15c)}$$

The corresponding representation for the laminar rotation (6.5b) reads

$$\psi = \sum_{n=0}^{\infty} \overset{(n)}{\psi}(x^\lambda)\, P_n(t) \qquad \textbf{(6.16)}$$

with the coefficients given by

$$\overset{(0)}{\psi} = \frac{1}{2}(\overset{(0)}{u}_{1,2} - \overset{(0)}{u}_{2,1}) \qquad \textbf{(6.17a)}$$

$$\overset{(n)}{\psi} = \frac{1}{2}(\overset{(n)}{u}_{1,2} - \overset{(n)}{u}_{2,1}),\ n \geq 1 \qquad \textbf{(6.17b)}$$

where, in the latter, the $\overset{(n)}{u}_\alpha$ are to be substituted from (6.12).

For the undetermined quantities introduced by the integration of the constitutive relations, we note that

$$\frac{1}{2h}\int_{-h}^{h} v_\alpha\, dz = \frac{1}{2}\int_{-1}^{1} u_\alpha\, dt = \overset{(0)}{u}_\alpha,\ \frac{1}{2h}\int_{-h}^{h} v_3\, dz = \frac{1}{2}\int_{-1}^{1} w\, dt = \overset{(0)}{w} \tag{6.18}$$

so that $\overset{(0)}{u}_\alpha$ and $\overset{(0)}{w}$, respectively, represent the mean laminar and mean transverse displacements: similarly $\overset{(0)}{\psi}$, and $\overset{(0)}{\omega}_\alpha$ respectively denote the mean laminar and mean transverse rotations.

We have now concluded the approximate integration of the three dimensional equations with respect to the thickness coordinate. Except for the undetermined mean values, all higher coefficients in the representations (6.7) and (6.9) have, to a first approximation, been expressed in terms of the elements of the original expansions (4.4) for the laminar stresses. This completes the determination of all field quantities in terms of the unknown laminar stress coefficients and mean displacements.

7. The Equations for the Unknown Functions

From the third constitutive relation, we shall derive the equations to be satisfied by the mean displacements and by the higher laminar stress coefficients. These together with equations (4.28) then constitute the full system of two-dimensional equations describing the three-dimensional problem.

We first introduce the representations for the displacements into the left hand side of (3.12c) while into the right hand side we introduce the corresponding expansions for the stress quantities. By rearranging we can then write each side as a series of Legendre polynomials, the orthogonality of which implies equality for the corresponding coefficients. This yields an infinite sequence of equations for the unknown quantities. Although the effect of the curvature results in the complete coupling of the system, so that mean displacement terms appear in the higher (i.e. $n > 2$) equations, we shall nevertheless refer to the first two as the equations for the determination of the mean displacements, and consider the set consisting of the remainder of the sequence as the system for the determination of the higher laminar coefficients.

Replacing the displacement terms in (3.12c) by their Legendre representations (6.7) and (6.9), we apply the recursion formula (4.1) and rearrange so that the left hand side takes the series form

$$\sum_{n=0}^{\infty}\Bigg[\frac{1}{2}\left(\overset{(n)}{u}_{\alpha|\beta} + \overset{(n)}{u}_{\beta|\alpha}\right) - b_{\alpha\beta}\overset{(n)}{w} + hb^{\lambda}_{\alpha|\beta}\left(\frac{n}{2n-1}\overset{(n-1)}{u}_{\lambda} + \frac{n+1}{2n+3}\overset{(n+1)}{u}_{\lambda}\right)$$
$$+ hb^{\lambda}_{\alpha}b_{\lambda\beta}\left(\frac{n}{2n-1}\overset{(n-1)}{w} + \frac{n+1}{2n+3}\overset{(n+1)}{w}\right)\Bigg]P_n \tag{7.1}$$

where the terms with index $(n-1)$ do not require any special attention for $n = 0$, since they are multiplied by n. Similarly by introducing the expansions (4.4) and (4.13) for the stress quantities, we again apply the recursion formula (4.1) and rearrange to obtain the corresponding Legendre series for the right hand side, namely

$$\frac{1}{E}\sum_{n=0}^{\infty}\Bigg\{(1+\nu)\Big[\overset{(n)}{S}_{\alpha\beta} + 2hH\Big(\frac{n}{2n-1}\overset{(n-1)}{S}_{\alpha\beta} + \frac{n+1}{2n+3}\overset{(n+1)}{S}_{\alpha\beta}\Big)$$

$$- 2hb_{\alpha\lambda}\Big(\frac{n}{2n-1}\overset{(n-1)}{S}{}^{\lambda}_{\beta} + \frac{n+1}{2n+3}\overset{(n+1)}{S}{}^{\lambda}_{\beta}\Big) - hb_{\beta\lambda}\Big(\frac{n}{2n-1}\overset{(n-1)}{S}{}^{\lambda}_{\alpha} + \frac{n+1}{2n+3}\overset{(n+1)}{S}{}^{\lambda}_{\alpha}\Big)\Big]$$

$$- g_{\alpha\beta}\Big[(\nu\overset{(n)}{S}{}^{\lambda}_{\lambda} + \nu_*\overset{(n)}{\sigma}) - \nu hb^{\lambda}_{\mu}\Big(\frac{n}{2n-1}\overset{(n-1)}{S}{}^{\mu}_{\lambda} + \frac{n+1}{2n+3}\overset{(n+1)}{S}{}^{\mu}_{\lambda}\Big)$$

$$+ 2hH\Big[\frac{n}{2n-1}(\nu\overset{(n-1)}{S}{}^{\lambda}_{\lambda} + \nu_*\overset{(n+1)}{\sigma}) + \frac{n+1}{2n+3}(\nu\overset{(n+1)}{S}{}^{\lambda}_{\lambda} + \nu_*\overset{(n+1)}{\sigma})\Big]\Big]$$

$$+ 2hb_{\alpha\beta}\Big[\frac{n}{2n-1}(\nu\overset{(n-1)}{S}{}^{\lambda}_{\lambda} + \nu_*\overset{(n+1)}{\sigma}) + \frac{n+1}{2n+3}(\nu\overset{(n+1)}{S}{}^{\lambda}_{\lambda} + \nu_*\overset{(n+1)}{\sigma})\Big]\Bigg\}P_n \tag{7.2}$$

The identification of (7.1) to (7.2) leads to the equations to be satisfied by the unknown functions.

The equations in which the mean laminar displacements predominate follow from the identification of the coefficients of P_0: if we recall formula (6.8) for $n = 1$, we see that, consistent with neglecting effects of order θ^2, the terms $h^2 b_{\alpha\lambda} b^{\lambda}_{\beta}\overset{(1)}{w}$ may be omitted and we obtain

$$E\Big[\frac{1}{2}(\overset{(0)}{u}_{\alpha|\beta} + \overset{(0)}{u}_{\beta|\alpha}) - b_{\alpha\beta}\overset{(0)}{w} + \frac{1}{3}hb^{\lambda}_{\alpha|\beta}\overset{(1)}{u}_{\lambda}\Big]$$

$$= (1+\nu)\Big[\overset{(0)}{S}_{\alpha\beta} + \frac{2}{3}hH\overset{(1)}{S}_{\alpha\beta} - \frac{2}{3}hb_{\alpha\lambda}\overset{(1)}{S}{}^{\lambda}_{\beta} - \frac{1}{3}hb_{\beta\lambda}\overset{(1)}{S}{}^{\lambda}_{\alpha}\Big]$$

$$- g_{\alpha\beta}\Big[(\nu\overset{(0)}{S}{}^{\lambda}_{\lambda} + \nu_*\overset{(0)}{\sigma}) - \frac{1}{3}\nu hb^{\lambda}_{\mu}\overset{(1)}{S}{}^{\mu}_{\lambda} + \frac{2}{3}hH(\nu\overset{(1)}{S}{}^{\lambda}_{\lambda} + \nu_*\overset{(1)}{\sigma})\Big]$$

$$+ \frac{2}{3}hb_{\alpha\beta}\Big[\nu\overset{(1)}{S}{}^{\lambda}_{\lambda} + \nu_*\overset{(1)}{\sigma}\Big] \tag{7.3}$$

Similarly, by identifying the coefficients of P_1, we obtain the equations in which the mean transverse displacement predominates, namely

$$E\Big[\frac{1}{2}(\overset{(1)}{u}_{\alpha|\beta} + \overset{(1)}{u}_{\beta|\alpha}) - b_{\alpha\beta}\overset{(1)}{w} + hb^{\lambda}_{\alpha|\beta}(\overset{(0)}{u}_{\lambda} + \frac{2}{5}\overset{(2)}{u}_{\lambda}) + hb^{\lambda}_{\alpha}b_{\lambda\beta}\overset{(0)}{w}\Big]$$

$$= (1+\nu)\Big[\overset{(1)}{S}_{\alpha\beta} + 2hH(\overset{(0)}{S}_{\alpha\beta} + \frac{2}{5}\overset{(2)}{S}_{\alpha\beta}) - 2hb_{\alpha\lambda}(\overset{(0)}{S}{}^{\lambda}_{\beta} + \frac{2}{5}\overset{(2)}{S}{}^{\lambda}_{\beta}) - hb_{\beta\lambda}(\overset{(0)}{S}{}^{\lambda}_{\alpha} + \frac{2}{5}\overset{(2)}{S}{}^{\lambda}_{\alpha})\Big]$$

$$- g_{\alpha\beta}\Big[(\nu\overset{(1)}{S}{}^{\lambda}_{\lambda} + \nu_*\overset{(1)}{\sigma}) + 2hH[(\nu\overset{(0)}{S}{}^{\lambda}_{\lambda} + \nu_*\overset{(0)}{\sigma}) + \frac{2}{5}(\nu\overset{(2)}{S}{}^{\lambda}_{\lambda} + \nu_*\overset{(2)}{\sigma})] - \nu hb^{\lambda}_{\mu}(\overset{(0)}{S}{}^{\mu}_{\lambda} + \frac{2}{5}\overset{(2)}{S}{}^{\mu}_{\lambda})$$

$$+ 2hb_{\alpha\beta}[(\nu\overset{(0)}{S}{}^{\lambda}_{\lambda} + \nu_*\overset{(0)}{\sigma}) + \frac{2}{5}(\nu\overset{(2)}{S}{}^{\lambda}_{\lambda} + \nu_*\overset{(2)}{\sigma})]\Big] \tag{7.4}$$

where we have omitted the terms $hb_{\alpha\lambda}b^{\lambda}_{\beta}\overset{(2)}{w}$, since by (6.8) they give rise to terms of order θ^2 compared with the corresponding terms on the right.

The identification of the higher coefficients yields the system of equations for the higher laminar coefficients. Again from (6.8), we observe that the last factor in the square bracket in the higher coefficients in (7.1) is neglegible compared with the corresponding terms on the right: omitting such terms, we have

$$E\left[\frac{1}{2}(\overset{(n)}{u}_{\alpha|\beta} + \overset{(n)}{u}_{\beta|\alpha}) - b_{\alpha\beta}\overset{(n)}{w} + hb^{\lambda}_{\alpha|\beta}\left(\frac{n}{2n-1}\overset{(n-1)}{u}_{\lambda} + \frac{n+1}{2n+3}\overset{(n+1)}{u}_{\lambda}\right)\right]$$

$$= (1+\nu)\left[\overset{(n)}{S}_{\alpha\beta} - 2hH\left(\frac{n}{2n-1}\overset{(n-1)}{S}_{\alpha\beta} + \frac{n+1}{2n+3}\overset{(n+1)}{S}{}^{\alpha}_{\beta}\right)\right.$$

$$\left. - 2hb_{\alpha\lambda}\left(\frac{n}{2n-1}\overset{(n+1)}{S}{}^{\lambda}_{\beta} + \frac{n+1}{2n+3}\overset{(n+1)}{S}{}^{\lambda}_{\beta}\right) - hb_{\beta\lambda}\left(\frac{n}{2n-1}\overset{(n-1)}{S}{}^{\lambda}_{\alpha} + \frac{n+1}{2n+3}\overset{(n+1)}{S}{}^{\lambda}_{\alpha}\right)\right]$$

$$- g_{\alpha\beta}\left[(\nu\overset{(n)}{S}{}^{\lambda}_{\lambda} + \nu_*\overset{(n)}{\sigma}) + 2hH[\frac{n}{2n-1}(\nu\overset{(n-1)}{S}{}^{\lambda}_{\lambda} + \nu_*\overset{(n-1)}{\sigma}) + \frac{n+1}{2n+3}(\nu\overset{(n+1)}{S}{}^{\lambda}_{\lambda} + \nu_*\overset{(n+1)}{\sigma})]\right.$$

$$\left. - \nu hb^{\lambda}_{\mu}\left(\frac{n}{2n-1}\overset{(n-1)}{S}{}^{\mu}_{\lambda} + \frac{n+1}{2n+3}\overset{(n+1)}{S}{}^{\mu}_{\lambda}\right)\right]$$

$$+ 2hb_{\alpha\beta}\left[\frac{n}{2n-1}(\nu\overset{(n-1)}{S}{}^{\lambda}_{\lambda} + \nu_*\overset{(n-1)}{\sigma}) + \frac{n+1}{2n+3}(\nu\overset{(n+1)}{S}{}^{\lambda}_{\lambda} + \nu_*\overset{(n+1)}{\sigma})\right], \quad n \geq 1 \tag{7.5}$$

which, with the $\overset{(n)}{u}_\alpha$ given by (6.12) and the $\overset{(n)}{w}$ given by (6.8), gives the completed sequence of equations.

The system of equations (4.28), (7.3), (7.4) and (7.5), when complemented by the edge conditions (5.11) and (5.12), constitute the approximate two-dimensional formulation of the boundary value problem, previously posed in three-dimensional terms by equations (3.10) to (3.12) with the boundary conditions (3.14) to (3.16). The approximation in the two-dimensional description is consistent with that already made in the formulation of the three-dimensional constitutive relations (3.12c).

8. Detachment of the Residual Problem

In order to examine equations (7.3) and (7.4) in more detail, it is necessary to have the higher displacement coefficients expressed in terms of the stresses and mean displacements. In particular, we must calculate $\overset{(1)}{u}_\alpha$ completely. Accordingly, we insert into formula (6.12a) the expressions for $\overset{(1)}{w}$, $\overset{(2)}{w}$ and $\overset{(3)}{w}$, as given by (6.8): if we also substitute for the $\overset{(n)}{S}_\alpha$ $(n = 0,1,2,3)$ from relations (4A.5) to (4A.8), then, after omitting terms of order θ^2 and some rearranging, we find

$$
\begin{aligned}
\overset{(1)}{u}_\alpha = {} & -h\,\overset{(0)}{w}|_\alpha - 2hb^\lambda_\alpha \overset{(0)}{u}_\lambda \\
& + 2\frac{h^2}{E}\Lambda(1+\nu)\Big[\Big(\frac{2}{5}\overset{(1)}{S}{}^\lambda_{\alpha|\lambda} - \frac{1}{35}\overset{(3)}{S}{}^\lambda_{\alpha|\lambda}\Big) + \frac{4}{105}hH\Big(\overset{(2)}{S}{}^\lambda_{\alpha|\lambda} - \frac{1}{6}\overset{(4)}{S}{}^\lambda_{\alpha|\lambda}\Big) \\
& \qquad - \frac{11}{105}hb^\lambda_\alpha\Big(\overset{(2)}{S}{}^\mu_{\lambda|\mu} - \frac{2}{33}\overset{(4)}{S}{}^\mu_{\lambda|\mu}\Big)\Big] \\
& + \frac{h^2}{E}\Bigg\{\Bigg[\frac{1}{15}\big(\Lambda_*\overset{(1)}{\sigma} - \nu_*\overset{(1)}{S}{}^\lambda_\lambda\big) - \frac{1}{35}\big(\Lambda_*\overset{(3)}{\sigma} - \nu_*\overset{(3)}{S}{}^\lambda_\lambda\big) \\
& \qquad + \frac{2}{15}hH\Big[\big(\Lambda_*\overset{(0)}{\sigma} - \nu_*\overset{(0)}{S}{}^\lambda_\lambda\big) + \frac{1}{7}\big(\Lambda_*\overset{(2)}{\sigma} - \nu_*\overset{(2)}{S}{}^\lambda_\lambda\big) - \frac{4}{21}\big(\Lambda_*\overset{(4)}{\sigma} - \nu_*\overset{(4)}{S}{}^\lambda_\lambda\big)\Big] \\
& \qquad + \frac{1}{15}\nu_* hb^\lambda_\mu\Big[\overset{(0)}{S}{}^\mu_\lambda + \frac{1}{7}\overset{(2)}{S}{}^\mu_\lambda - \frac{4}{21}\overset{(4)}{S}{}^\mu_\lambda\Big]\Bigg]_{|\alpha} \\
& \qquad - \frac{2}{15}hb^\lambda_\alpha\Big[\big(\Lambda_*\overset{(0)}{\sigma} - \nu_*\overset{(0)}{S}{}^\mu_\mu\big) - \frac{2}{7}\big(\Lambda_*\overset{(2)}{\sigma} - \nu_*\overset{(2)}{S}{}^\mu_\mu\big) + \frac{1}{21}\big(\Lambda_*\overset{(4)}{\sigma} - \nu_*\overset{(4)}{S}{}^\mu_\mu\big)\Big]_{|\lambda}\Bigg\}
\end{aligned}
\tag{8.1}
$$

Insofar as its introduction into (7.3) is concerned, many terms in (8.1) are negligible: however, the full formula is necessary in (7.4). In the latter equation there also occur the coefficients $\overset{(1)}{w}$ and $\overset{(2)}{u}_\alpha$: however from the combinations in which they occur, we see that it is not necessary to have the complete expression for these quantities. Thus an inspection of (6.8) shows that for its insertion into (7.4) we may take

$$\overset{(1)}{w} \simeq \frac{h}{E}\left[\left(\Lambda_* \overset{(0)}{\sigma} - \nu_* \overset{(0)}{S}{}^{\lambda}_{\lambda}\right) - \frac{1}{5}\left(\Lambda_* \overset{(2)}{\sigma} - \nu_* \overset{(2)}{S}{}^{\lambda}_{\lambda}\right)\right] \tag{8.2}$$

The expression for $\overset{(2)}{u}_\alpha$ may be calculated in the same manner as was done for $\overset{(1)}{u}_\alpha$: however for its use in (7.4) it is consistent with neglecting effects of order θ^2 to use the simplified formula

$$\begin{aligned}\overset{(2)}{u}_\alpha \simeq{}& \frac{h}{E}\, 2\Lambda(1+\nu)\left[\frac{2}{21}\, h \overset{(2)}{S}{}^{\lambda}_{\alpha|\lambda} - \frac{1}{63}\, h \overset{(4)}{S}{}^{\lambda}_{\alpha|\lambda}\right] \\ &- \frac{1}{3}\frac{h^2}{E}\left[\left(\Lambda_* \overset{(0)}{\sigma} - \nu_* \overset{(0)}{S}{}^{\lambda}_{\lambda}\right) - \frac{2}{7}\left(\Lambda_* \overset{(2)}{\sigma} - \nu_* \overset{(2)}{S}{}^{\lambda}_{\lambda}\right) + \frac{1}{21}\left(\Lambda_* \overset{(4)}{\sigma} - \nu_* \overset{(4)}{S}{}^{\lambda}_{\lambda}\right)\right]_{|\alpha}\end{aligned} \tag{8.3}$$

We now introduce expression (8.1) into equation (7.3) and, neglecting terms of order θ^2, we impose an arrangement that leaves only mean displacement terms on the left: there follows

$$\begin{aligned}&E\left[\frac{1}{2}\left(\overset{(0)}{u}_{\alpha|\beta} + \overset{(0)}{u}_{\beta|\alpha}\right) - b_{\alpha\beta}\overset{(0)}{w} - \frac{1}{3} h^2 b^{\lambda}_{\alpha|\beta}\, \overset{(0)}{w}|_{\lambda}\right] \\ &= (1+\nu)\left[\overset{(0)}{S}_{\alpha\beta} + \frac{2}{3} hH \overset{(1)}{S}_{\alpha\beta} - \frac{2}{3} hb_{\alpha\lambda} \overset{(1)}{S}{}^{\lambda}_{\beta} - \frac{1}{3} hb_{\beta\lambda} \overset{(1)}{S}{}^{\lambda}_{\alpha}\right] + \frac{2}{3} hb_{\alpha\beta}\left[\nu \overset{(1)}{S}{}^{\lambda}_{\lambda} + \nu_* \overset{(1)}{\sigma}\right] \\ &- g_{\alpha\beta}\left[\left(\nu \overset{(0)}{S}{}^{\lambda}_{\lambda} + \nu_* \overset{(0)}{\sigma}\right) + \frac{2}{3} hH\left(\nu \overset{(1)}{S}{}^{\lambda}_{\lambda} + \nu_* \overset{(1)}{\sigma}\right) - \frac{1}{3}\nu h b^{\lambda}_{\mu} \overset{(1)}{S}{}^{\mu}_{\lambda}\right] \\ &- \frac{4}{15} h^3 b^{\lambda}_{\alpha|\beta}\left[\Lambda(1+\nu)\left(\overset{(1)}{S}{}^{\mu}_{\lambda|\mu} - \frac{1}{14} \overset{(3)}{S}{}^{\mu}_{\lambda|\mu}\right) + \frac{1}{12}\left(\Lambda_* \overset{(1)}{\sigma} - \nu_* \overset{(1)}{S}{}^{\mu}_{\mu}\right)_{|\lambda} - \frac{1}{28}\left(\Lambda_* \overset{(3)}{\sigma} - \nu_* \overset{(3)}{S}{}^{\mu}_{\mu}\right)_{|\lambda}\right]\end{aligned} \tag{8.4}$$

If we also introduce formulae (8.1), (8.2) and (8.3) into (7.4) and make a similar transposition, then, after considerable rearrangement, we obtain

$$-Eh[\overset{(0)}{w}|_{\alpha\beta} + b^{\lambda}_{\alpha}\overset{(0)}{u}_{\lambda|\beta} + b^{\lambda}_{\beta}\overset{(0)}{u}_{\lambda|\alpha} + b^{\lambda}_{\alpha|\beta}\overset{(0)}{u}_{\lambda} - b_{\alpha\lambda}b^{\lambda}_{\beta}\overset{(0)}{w}]$$

$$= (1+\nu)\left[\overset{(1)}{S}_{\alpha\beta} + 2hH(\overset{(0)}{S}_{\alpha\beta} + \tfrac{2}{5}\overset{(2)}{S}_{\alpha\beta}) - 2hb_{\alpha\lambda}(\overset{(0)}{S}{}^{\lambda}_{\beta} + \tfrac{2}{5}\overset{(2)}{S}{}^{\lambda}_{\beta}) - hb_{\beta\lambda}(\overset{(0)}{S}{}^{\lambda}_{\alpha} + \tfrac{2}{5}\overset{(2)}{S}{}^{\lambda}_{\alpha})\right]$$

$$+ hb_{\alpha\beta}\left[\left((\Lambda_* + 2\nu_*)\overset{(0)}{\sigma} + (2\nu - \nu_*)\overset{(0)}{S}{}^{\lambda}_{\lambda}\right) - \tfrac{1}{5}\left((\Lambda_* - 4\nu_*)\overset{(2)}{\sigma} - (4\nu + \nu_*)\overset{(2)}{S}{}^{\lambda}_{\lambda}\right)\right]$$

$$- g_{\alpha\beta}\left[(\nu\overset{(1)}{S}{}^{\lambda}_{\lambda} + \nu_*\overset{(1)}{\sigma}) + 2hH[(\nu\overset{(0)}{S}{}^{\lambda}_{\lambda} + \nu_*\overset{(0)}{\sigma}) + \tfrac{2}{5}(\nu\overset{(2)}{S}{}^{\lambda}_{\lambda} + \nu_*\overset{(2)}{\sigma})] - \nu hb^{\lambda}_{\mu}(\overset{(0)}{S}{}^{\mu}_{\lambda} + \tfrac{2}{5}\overset{(2)}{S}{}^{\mu}_{\lambda})\right]$$

$$- \Lambda(1+\nu)\Big[\tfrac{2}{5}h^2(\overset{(1)}{S}{}^{\lambda}_{\alpha|\lambda\beta} + \overset{(1)}{S}{}^{\lambda}_{\beta|\lambda\alpha}) - \tfrac{1}{35}h^2(\overset{(3)}{S}{}^{\lambda}_{\alpha|\lambda\beta} + \overset{(3)}{S}{}^{\lambda}_{\beta|\lambda\alpha})$$

$$+ \tfrac{8}{105}h^3[(H\overset{(2)}{S}{}^{\lambda}_{\alpha|\lambda})_{|\beta} - \tfrac{1}{6}(H\overset{(4)}{S}{}^{\lambda}_{\alpha|\lambda})_{|\beta}]$$

$$- \tfrac{22}{105}h^3[(b^{\lambda}_{\alpha}\overset{(2)}{S}{}^{\mu}_{\lambda|\mu})_{|\beta} - \tfrac{2}{33}(b^{\lambda}_{\alpha}\overset{(4)}{S}{}^{\mu}_{\lambda|\mu})_{|\beta}]$$

$$+ \tfrac{8}{105}h^3 b^{\lambda}_{\alpha|\beta}[\overset{(2)}{S}{}^{\mu}_{\lambda|\mu} - \tfrac{1}{6}\overset{(4)}{S}{}^{\mu}_{\lambda|\mu}]\Big]$$

$$- h^2\Big[\tfrac{1}{15}(\Lambda_*\overset{(1)}{\sigma} - \nu_*\overset{(1)}{S}{}^{\lambda}_{\lambda}) - \tfrac{1}{35}(\Lambda_*\overset{(3)}{\sigma} - \nu_*\overset{(3)}{S}{}^{\lambda}_{\lambda}) + \tfrac{1}{15}\nu_* hb^{\lambda}_{\mu}[\overset{(0)}{S}{}^{\mu}_{\lambda} + \tfrac{1}{7}\overset{(2)}{S}{}^{\mu}_{\lambda} - \tfrac{4}{21}\overset{(4)}{S}{}^{\mu}_{\lambda}]$$

$$+ \tfrac{2}{15}hH[(\Lambda_*\overset{(0)}{\sigma} - \nu_*\overset{(0)}{S}{}^{\lambda}_{\lambda}) + \tfrac{1}{7}(\Lambda_*\overset{(2)}{\sigma} - \nu_*\overset{(2)}{S}{}^{\lambda}_{\lambda}) - \tfrac{4}{21}(\Lambda_*\overset{(4)}{\sigma} - \nu_*\overset{(4)}{S}{}^{\lambda}_{\lambda})]\Big]_{|\alpha\beta}$$

$$- \tfrac{2}{15}h^3 b^{\lambda}_{\alpha}[(\Lambda_*\overset{(0)}{\sigma} - \nu_*\overset{(0)}{S}{}^{\mu}_{\mu}) - \tfrac{2}{7}(\Lambda_*\overset{(2)}{\sigma} - \nu_*\overset{(2)}{S}{}^{\mu}_{\mu}) + \tfrac{1}{21}(\Lambda_*\overset{(4)}{\sigma} - \nu_*\overset{(4)}{S}{}^{\mu}_{\mu})]_{|\lambda\beta}$$

$$- \tfrac{4}{15}h^3 b^{\lambda}_{\alpha|\beta}[(\Lambda_*\overset{(0)}{\sigma} - \nu_*\overset{(0)}{S}{}^{\mu}_{\mu}) - \tfrac{2}{7}(\Lambda_*\overset{(2)}{\sigma} - \nu_*\overset{(2)}{S}{}^{\mu}_{\mu}) + \tfrac{1}{21}(\Lambda_*\overset{(4)}{\sigma} - \nu_*\overset{(4)}{S}{}^{\mu}_{\mu})]_{|\lambda} \quad (8.5)$$

where we have consistently retained the unsymmetric form on the right hand side by noting the symmetry conditions (4.6).

Equations (8.4) and (8.5) indicate how the residual components of the mean displacements are to be defined. Returning to relations (4A.9) to (4A.13), we decompose the transverse normal stress coefficients into their principal and residual parts by setting

$$\overset{(n)}{\sigma} = \overset{(n)}{\sigma}{}^{P} + \overset{(n)}{\sigma}{}^{R} \tag{8.6}$$

where by inspection we have

$$\overset{(0)}{\sigma}{}^{P} = \frac{1}{2} p^* + \frac{1}{3} h b^{\lambda}_{\mu} \overset{(1)}{S}{}^{\mu}_{\lambda} + \frac{4}{15} h^3 (b^{\lambda}_{\rho} \overset{(1)}{S}{}^{\mu\rho}|_{\mu})_{|\lambda}, \tag{8.7a}$$

$$\overset{(1)}{\sigma}{}^{P} = \frac{3}{5} p + \frac{1}{5} h b^{\lambda}_{\mu} \overset{(0)}{S}{}^{\mu}_{\lambda}, \tag{8.7b}$$

$$\overset{(2)}{\sigma}{}^{P} = -\frac{1}{3} h b^{\lambda}_{\mu} \overset{(1)}{S}{}^{\mu}_{\lambda} - \frac{8}{63} h^3 (b^{\lambda}_{\rho} \overset{(1)}{S}{}^{\mu\rho}|_{\mu})_{|\lambda}, \tag{8.7c}$$

$$\overset{(3)}{\sigma}{}^{P} = -\frac{1}{10} p - \frac{1}{5} h b^{\lambda}_{\mu} \overset{(0)}{S}{}^{\mu}_{\lambda}, \tag{8.7d}$$

$$\overset{(4)}{\sigma}{}^{P} = -\frac{4}{105} h^3 (b^{\lambda}_{\rho} \overset{(1)}{S}{}^{\mu\rho}|_{\mu})_{|\lambda} \tag{8.7e}$$

$$\overset{(n)}{\sigma}{}^{P} = 0, \quad n \geq 5 \tag{8.7f}$$

and the $\overset{(n)}{\sigma}{}^{R}$ are the complementary expressions involving only the residual laminar stresses $\{ \overset{(n)}{S}{}^{\lambda\mu}, n \geq 2\}$. If we now connect the decomposition (8.6) with equations (8.4) and (8.5), we see that the appropriate definition of the residual components of the mean displacements is through the pair of determining equations

$$E[\frac{1}{2}(\overset{(R)}{U}_{\alpha|\beta} + \overset{(R)}{U}_{\beta|\alpha}) - b_{\alpha\beta} \overset{(R)}{W} - \frac{1}{3} h^2 b^{\lambda}_{\alpha|\beta} \overset{(R)}{W}_{|\lambda}]$$

$$= -\nu_* g_{\alpha\beta} [\overset{(0)}{\sigma}{}^{R} + \frac{2}{3} hH \overset{(1)}{\sigma}{}^{R}] + \frac{2}{3} \nu_* h b_{\alpha\beta} \overset{(0)}{\sigma}{}^{R}$$

$$+ \frac{2}{15} h^2 b^{\lambda}_{\alpha|\beta} [\Lambda (1+\nu) \frac{1}{7} h \overset{(3)}{S}{}^{\mu}_{\lambda|\mu} - \frac{1}{6} \Lambda_* h \overset{(1)}{\sigma}{}^{R}{}_{|\lambda} + \frac{1}{14} (\Lambda_* h \overset{(3)}{\sigma}{}^{R}{}_{|\lambda} - \nu_* h \overset{(3)}{S}{}^{\mu}_{\mu|\lambda})] \tag{8.8}$$

and

$$
\begin{aligned}
&- Eh[\overset{(R)}{W}|_{\alpha\beta} + b^{\lambda}_{\alpha}\overset{(R)}{U}_{\lambda|\beta} + b^{\lambda}_{\beta}\overset{(R)}{U}_{\lambda|\alpha} + b^{\lambda}_{\alpha|\beta}\overset{(R)}{U}_{\lambda} - b_{\alpha\lambda} b^{\lambda}_{\beta}\overset{(R)}{W}] \\
&= (1+\nu)\,[\tfrac{4}{5}\, hH \overset{(2)}{S}_{\alpha|\beta} - \tfrac{4}{5}\, hb_{\alpha\lambda}\overset{(2)}{S}{}^{\lambda}_{\beta} - \tfrac{2}{5}\, hb_{\beta\lambda}\overset{(2)}{S}{}^{\mu}_{\alpha}] \\
&- g_{\alpha\beta}\left[\nu_*\overset{(1)}{\sigma}{}^{R} + 2\mathrm{h}H[\nu_*\overset{(0)}{\sigma}{}^{R} + \tfrac{2}{5}(\nu\overset{(2)}{S}{}^{\lambda}_{\lambda} + \nu_*\overset{(2)}{\sigma}{}^{R})] - \tfrac{2}{5}\,\nu h b^{\lambda}_{\mu}\overset{(2)}{S}{}^{\mu}_{\lambda}\right] \\
&+ hb_{\alpha\beta}\left[(\Lambda_* + 2\nu_*)\overset{(0)}{\sigma}{}^{R} - \tfrac{1}{5}[(\Lambda_* - 4\nu_*)\overset{(2)}{\sigma}{}^{R} - (4\nu + \nu_*)\overset{(2)}{S}{}^{\lambda}_{\lambda}\right] \\
&+ \Lambda(1+\nu)\left[\tfrac{1}{35}\, h^2(\overset{(3)}{S}{}^{\lambda}_{\alpha|\lambda\beta} + \overset{(3)}{S}{}^{\lambda}_{\beta|\lambda\alpha}) - \tfrac{8}{105}\, h^3[(H\overset{(2)}{S}{}^{\lambda}_{\alpha|\lambda})_{|\beta} - \tfrac{1}{6}(H\overset{(4)}{S}{}^{\lambda}_{\alpha|\lambda})|_{\beta}]\right. \\
&\qquad + \tfrac{22}{105}\, h^3[(b^{\lambda}_{\alpha}\overset{(2)}{S}{}^{\mu}_{\lambda|\mu})_{|\beta} - \tfrac{2}{33}(b^{\lambda}_{\alpha}\overset{(4)}{S}{}^{\mu}_{\lambda|\mu})_{|\beta}] - \tfrac{8}{105}\, h^3 b^{\lambda}_{\alpha|\beta}[\overset{(2)}{S}{}^{\mu}_{\lambda|\mu} - \tfrac{1}{6}\overset{(4)}{S}{}^{\mu}_{\lambda|\mu}]\Big] \\
&- h^2\Big[\tfrac{1}{15}\Lambda_*\overset{(1)}{\sigma}{}^{R} - \tfrac{1}{35}(\Lambda_*\overset{(3)}{\sigma}{}^{R} - \nu_*\overset{(3)}{S}{}^{\lambda}_{\lambda}) + \tfrac{1}{15}\nu_* h b^{\lambda}_{\mu}(\tfrac{1}{7}\overset{(2)}{S}{}^{\mu}_{\lambda} - \tfrac{4}{21}\overset{(4)}{S}{}^{\mu}_{\lambda}) \\
&\qquad + \tfrac{2}{15}\, hH[\Lambda_*\overset{(2)}{\sigma}{}^{R} + \tfrac{1}{7}(\Lambda_*\overset{(2)}{\sigma}{}^{R} - \nu_*\overset{(2)}{S}{}^{\lambda}_{\lambda}) - \tfrac{4}{21}(\Lambda_*\overset{(4)}{\sigma}{}^{R} - \nu_*\overset{(4)}{S}{}^{\lambda}_{\lambda})]\Big]_{|\alpha\beta} \\
&\qquad - \tfrac{2}{15}\, h^3 b^{\lambda}_{\alpha}[\Lambda_*\overset{(0)}{\sigma}{}^{R} - \tfrac{2}{7}(\Lambda_*\overset{(2)}{\sigma}{}^{R} - \nu_*\overset{(2)}{S}{}^{\mu}_{\mu}) + \tfrac{1}{21}(\Lambda_*\overset{(4)}{\sigma}{}^{R} - \nu_*\overset{(4)}{S}{}^{\mu}_{\mu})]_{|\lambda\beta} \\
&\qquad - \tfrac{4}{15}\, h^3 b^{\lambda}_{\alpha|\beta}[\Lambda_*\overset{(0)}{\sigma}{}^{R} - \tfrac{2}{7}(\Lambda_*\overset{(2)}{\sigma}{}^{R} - \nu_*\overset{(2)}{S}{}^{\mu}_{\mu}) + \tfrac{1}{21}(\Lambda_*\overset{(4)}{\sigma}{}^{R} - \nu_*\overset{(4)}{S}{}^{\mu}_{\mu})]_{|\lambda} \qquad \mathbf{(8.9)}
\end{aligned}
$$

Equations (8.8), (8.9) and (7.5), when complemented by the edge conditions (5.12) constitute the boundary value problem for the residual effects. The analysis of this residual problem presumes the prior solution of the principal problem since the principal effects appear as inhomogeneous terms both in equations (7.5) and in the edge conditions (5.12b).

The subsequent treatment of the equations for the residual effects would follow the pattern set for the residual problems in the stretching and bending of flat plates. We do not discuss it further.

9. The Principal Problem

Decomposing the mean displacements in accordance with the resolution

$$u_\alpha = \overset{(0)}{U}_\alpha + \overset{(R)}{U}_\alpha, \quad w = \overset{(0)}{W} + \overset{(R)}{W} \tag{9.2a,b}$$

the equation for the determination of the principal components U_α and W then follow from the subtraction of equation (8.8) and (8.9) respectively from (8.4) and (8.5). Taking the difference between (8.4) and (8.8), we substitute formulae (8.7) for the coefficients $\overset{(0)}{\sigma}{}^{P}$, $\overset{(1)}{\sigma}{}^{P}$ and $\overset{(3)}{\sigma}{}^{P}$ and on omitting effects of order θ^2 we obtain

$$E[\tfrac{1}{2}(U_{\alpha|\beta} + U_{\beta|\alpha}) - b_{\alpha\beta} W - \tfrac{1}{3} h^2 b^\lambda_{\alpha|\beta} W|_\lambda]$$

$$= (1+\nu)\,[\overset{(0)}{S}_{\alpha\beta} + \tfrac{2}{3} hH \overset{(1)}{S}_{\alpha\beta} - \tfrac{2}{3} h b_{\alpha\lambda} \overset{(1)}{S}{}^{\lambda}_{\beta} - \tfrac{1}{3} h b_{\beta\lambda} \overset{(1)}{S}{}^{\lambda}_{\alpha}]$$

$$- g_{\alpha\beta}[\nu(\overset{(0)}{S}{}^{\lambda}_{\lambda} + \tfrac{2}{3} hH \overset{(1)}{S}{}^{\lambda}_{\lambda}) - \tfrac{1}{3}(\nu - \nu_*) h b^\lambda_\mu \overset{(1)}{S}{}^{\mu}_{\lambda} + \tfrac{4}{15} \nu_* h^3 (b^\lambda_\rho \overset{(1)}{S}{}^{\mu\rho}|_\mu)_{|\lambda}]$$

$$+ \tfrac{2}{3} \nu h b_{\alpha\beta} \overset{(1)}{S}{}^{\lambda}_{\lambda} - \tfrac{4}{15} h^3 b^\lambda_{\alpha|\beta}[\Lambda(1+\nu) \overset{(1)}{S}{}^{\mu}_{\lambda|\mu} - \tfrac{1}{12} \nu_* \overset{(1)}{S}{}^{\mu}_{\mu|\lambda}]$$

$$- \tfrac{1}{2} \nu_* [g_{\alpha\beta}(p^* + \tfrac{4}{5} hHp) - \tfrac{4}{5} h b_{\alpha\beta} p] - \tfrac{1}{70} \Lambda_* h^3 b^\lambda_{\alpha|\beta} p|_\lambda \tag{9.2}$$

Similarly, after subtracting (8.9) from (8.5), the introduction of $\overset{(0)}{\sigma}{}^{P}$, $\overset{(1)}{\sigma}{}^{P}$, $\overset{(2)}{\sigma}{}^{P}$ and $\overset{(4)}{\sigma}{}^{P}$ from (8.7) yields

$$- Eh[W|_{\alpha\beta} + b^\lambda_\alpha U_{\lambda|\beta} + b^\lambda_\beta U_{\lambda|\alpha} + b^\lambda_{\alpha|\beta} U_\lambda - b_{\alpha\lambda} b^\lambda_\beta W]$$

$$= (1+\nu)\,[\overset{(1)}{S}_{\alpha\beta} + 2hH \overset{(0)}{S}_{\alpha\beta} - 2h b_{\alpha\lambda} \overset{(0)}{S}{}^{\lambda}_{\beta} - h b_{\beta\lambda} \overset{(0)}{S}{}^{\lambda}_{\alpha}] + (2\nu - \nu_*) h b_{\alpha\beta} \overset{(0)}{S}{}^{\lambda}_{\lambda}$$

$$- g_{\alpha\beta}[\nu(\overset{(1)}{S}{}^{\lambda}_{\lambda} + 2hH \overset{(0)}{S}{}^{\lambda}_{\lambda}) - \tfrac{1}{5}(5\nu - \nu_*) h b^\lambda_\mu \overset{(0)}{S}{}^{\mu}_{\lambda}] - \Lambda(1+\nu) \tfrac{2}{5} h^2 [\overset{(1)}{S}{}^{\lambda}_{\alpha|\lambda\beta} + \overset{(1)}{S}{}^{\lambda}_{\beta|\lambda\alpha}]$$

$$+ \tfrac{1}{15} h^2 [\nu_*(\overset{(1)}{S}{}^{\lambda}_{\lambda} + 2hH \overset{(0)}{S}{}^{\lambda}_{\lambda}) - (\nu_* + \tfrac{2}{7}\Lambda_*) h b^\lambda_\mu \overset{(0)}{S}{}^{\mu}_{\lambda}]_{|\alpha\beta} + \tfrac{2}{15} \nu_* h^3 [b^\lambda_\alpha \overset{(0)}{S}{}^{\mu}_{\mu|\lambda\beta} + 2 b^\lambda_{\alpha|\beta} \overset{(0)}{S}{}^{\mu}_{\mu|\lambda}]$$

$$- \nu_* g_{\alpha\beta}(\tfrac{3}{5} p + hHp^*) + \tfrac{1}{2}(\Lambda_* + 2\nu_*) h b_{\alpha\beta} p^* - \Lambda_* h^2 (\tfrac{3}{70} p + \tfrac{1}{15} hHp^*)_{|\alpha\beta}$$

$$- \tfrac{1}{15} \Lambda_* (h^3 b^\lambda_\alpha p^*|_{\lambda\beta} + 2h^3 b^\lambda_{\alpha|\beta} p^*|_\lambda) \tag{9.3}$$

where again we have neglected effects or order θ^2. Equations (9.2) and (9.3) are to be considered in conjunction with the equilibrium equations (4.28), and the edge conditions (5.11).

The trace of the system of equations (9.2) yields the relation

$$
\begin{aligned}
E[U^{\lambda}|_{\lambda} - 2HW - \tfrac{1}{3}h^2 b^{\mu\lambda}|_{\mu}\, W|_{\lambda}]
&= (1-\nu)\, \overset{(0)}{S}{}^{\lambda}_{\lambda} + (1+\nu)\tfrac{2}{3} hH\, \overset{(1)}{S}{}^{\lambda}_{\lambda} - [1 + \tfrac{1}{3}(\nu+2\nu_*)]\, h b^{\lambda}_{\mu}\, \overset{(1)}{S}{}^{\mu}_{\lambda} \\
&\quad - \tfrac{4}{15}h^3 \Big[b^{\rho\lambda}|_{\rho} [\Lambda(1+\nu)\, \overset{(1)}{S}{}^{\mu}_{\lambda|\mu} - \tfrac{1}{12}\nu_*\, \overset{(1)}{S}{}^{\mu}_{\mu|\lambda}] + 2\nu_* (b^{\lambda}_{\rho}\, \overset{(1)}{S}{}^{\mu\rho}|_{\mu})|_{\lambda} \Big] \\
&\quad - \nu_* p - \tfrac{1}{70}\Lambda_* h^3 b^{\rho\lambda}|_{\rho}\, p|_{\lambda} \qquad (9.4)
\end{aligned}
$$

A second relation is obtained by considering the contravariant form of (9.2) and taking the repeated covariant derivative with respect to the indicated variables (first with respect to x^{α} and then with respect to x^{β}): if we then sum over the repeated indices and utilize equation (4.28a) in the resulting equation, we find

$$
\begin{aligned}
E[U^{\alpha}|^{\beta}_{\alpha\beta} - (b^{\alpha}_{\beta}W)|^{\beta}_{\alpha} - \tfrac{1}{3}h^2 (b^{\alpha\lambda}|_{\beta}\, W|_{\lambda})|^{\beta}_{\alpha}]
&= (1+\nu)[\tfrac{1}{3}h(b^{\beta}_{\lambda}\, \overset{(1)}{S}{}^{\alpha\lambda}|_{\alpha})|_{\beta} + \tfrac{2}{3}h(H\, \overset{(1)}{S}{}^{\alpha}_{\beta})|^{\beta}_{\alpha} - h(b^{\alpha}_{\lambda}\, \overset{(1)}{S}{}^{\lambda}_{\beta})|^{\beta}_{\alpha}] \\
&\quad - \nu[\overset{(0)}{S}{}^{\lambda}_{\lambda}|^{\alpha}_{\alpha} + \tfrac{2}{3}h(H\, \overset{(1)}{S}{}^{\lambda}_{\lambda})|^{\alpha}_{\alpha}] + \tfrac{1}{3}(\nu-\nu_*)h(b^{\lambda}_{\mu}\, \overset{(1)}{S}{}^{\mu}_{\lambda})|^{\alpha}_{\alpha} + \tfrac{2}{3}\nu h(b^{\alpha}_{\beta}\, \overset{(1)}{S}{}^{\lambda}_{\lambda})|^{\beta}_{\alpha} \\
&\quad - \tfrac{4}{15}h^3[\Lambda(1+\nu)(b^{\alpha\lambda}|_{\beta}\, \overset{(1)}{S}{}^{\mu}_{\lambda|\mu})|^{\beta}_{\alpha} - \tfrac{1}{12}\nu_*(b^{\alpha\lambda}|_{\beta}\, \overset{(1)}{S}{}^{\mu}_{\mu|\lambda})|^{\beta}_{\alpha} + \nu_*(b^{\lambda}_{\rho}\, \overset{(1)}{S}{}^{\mu\rho}|_{\mu})|^{\alpha}_{\lambda\alpha}] \\
&\quad - \tfrac{1}{2}\nu_*(p_* + \tfrac{4}{5}hHp)|^{\alpha}_{\alpha} + \tfrac{2}{5}\nu_* h(b^{\alpha}_{\beta}p)|^{\beta}_{\alpha} - \tfrac{1}{70}\Lambda_* h^3 (b^{\alpha\lambda}|_{\beta}\, p|_{\lambda})|^{\beta}_{\alpha} \qquad (9.5)
\end{aligned}
$$

We now apply the Laplacian operator to equation (9.4) and, on subtracting (9.5) from the result, we obtain

$$
\begin{aligned}
&-E\Big[2(HW)|^{\alpha}_{\alpha} - (b^{\alpha}_{\beta}W)|^{\beta}_{\alpha} + \tfrac{1}{3}h^2[(b^{\lambda\mu}|_{\mu}\, W|_{\lambda})|^{\alpha}_{\alpha} - (b^{\alpha\lambda}|_{\beta}\, W|_{\lambda})|^{\beta}_{\alpha}] \Big] \\
&= \overset{(0)}{S}{}^{\lambda}_{\lambda}|^{\alpha}_{\alpha} + (1+2\nu)\tfrac{2}{3}h(H\, \overset{(1)}{S}{}^{\lambda}_{\lambda})|^{\alpha}_{\alpha} - [1 + \tfrac{1}{3}(2\nu+\nu_*)]h(b^{\lambda}_{\mu}\, \overset{(1)}{S}{}^{\mu}_{\lambda})|^{\alpha}_{\alpha}
\end{aligned}
$$

$$-\frac{1}{3}(1+\nu)h[(b^\beta_\lambda \overset{(1)}{S}{}^{\alpha\lambda}|_\alpha)|_\beta + 2(H\overset{(1)}{S}{}^\alpha_\beta)|^\beta_\alpha - 3(b^\alpha_\lambda \overset{(1)}{S}{}^\lambda_\beta)|^\beta_\alpha] - \frac{2}{3}\nu h(b^\alpha_\beta \overset{(1)}{S}{}^\lambda_\lambda)|^\beta_\alpha$$

$$-\frac{4}{15}h^3\Big[\Lambda(1+\nu)[(b^{\rho\lambda}|_\rho \overset{(1)}{S}{}^\mu_{\lambda|\mu})|^\alpha_\alpha - (b^{\alpha\lambda}|_\beta \overset{(1)}{S}{}^\mu_{\lambda|\mu})|^\beta_\alpha] + \nu_*(b^\lambda_\rho \overset{(1)}{S}{}^{\mu\rho}|_\mu)|^\alpha_{\lambda\alpha}$$

$$\frac{1}{12}\nu_*[(b^{\rho\lambda}|_\rho \overset{(1)}{S}{}^{\mu\lambda}|_\mu)|^\alpha_\alpha - (b^{\alpha\lambda}|_\beta \overset{(1)}{S}{}^\mu_{\lambda|\mu})|^\beta_\alpha]\Big]$$

$$-\frac{1}{2}\nu_*(p^* - \frac{4}{5}hHp)|^\alpha_\alpha - \frac{2}{5}\nu_* h(b^\alpha_\beta p)|^\beta_\alpha - \frac{1}{70}\Lambda_*[(b^{\rho\lambda}|_\rho p|_\lambda)|^\alpha_\alpha - (b^{\alpha\lambda}|_\beta p|_\lambda)|^\beta_\alpha] \tag{9.6}$$

which is the analog for the shell configuration of the well known compatibility equation for the generalized plane stress theory of the stretching of flat plates.

We now derive the corresponding equations resulting from second set of relations (9.3). The trace of this latter system yields the relation

$$-Eh[W|^\lambda_\lambda + 2b^\mu_\lambda U^\lambda|_\mu + b^{\mu\lambda}|_\mu U_\lambda - b^\lambda_\mu b^\mu_\lambda W]$$

$$= (1-\nu)\overset{(1)}{S}{}^\lambda_\lambda + (1+\nu-\nu_*)2hH\overset{(0)}{S}{}^\lambda_\lambda + [12\Lambda(1+\nu) - (15+5\nu+2\nu_*)]\frac{1}{5}hb^\lambda_\mu \overset{(0)}{S}{}^\mu_\lambda$$

$$+\frac{6}{5}[\Lambda(1+\nu) - \nu_*]p - \frac{2}{105}\Lambda_* h^3(b^\lambda_\mu \overset{(0)}{S}{}^\mu_\lambda)|^\alpha_\alpha$$

$$+\frac{1}{15}\nu_* h^2\Big[[\overset{(1)}{S}{}^\lambda_\lambda + 2hH\overset{(0)}{S}{}^\lambda_\lambda - hb^\lambda_\mu \overset{(0)}{S}{}^\mu_\lambda]|^\alpha_\alpha + 2h[b^\lambda_\alpha \overset{(0)}{S}{}^\mu_\mu|^\alpha_\lambda + 2b^{\alpha\lambda}|_\alpha \overset{(0)}{S}{}^\mu_{\mu|\lambda}]\Big]$$

$$+\Lambda_*\Big[hHp^* - h^2[(\frac{3}{70}p + \frac{1}{15}hHp^*)|^\alpha_\alpha + \frac{1}{15}h(b^\lambda_\alpha p^*|^\alpha_\lambda + 2b^{\alpha\lambda}|_\alpha p^*|_\lambda)]\Big] \tag{9.7}$$

As before, we deduce a second relation from taking the repeated covariant derivative, with respect to the indicated variables, of the contravariant form of (9.3): after summing over the repeated indices, we multiply by h^2 and utilizing equation (4.28b), we rearrange and obtain

$$-Eh^3[W|^{\alpha\beta}_{\alpha\beta} + 2(b^\alpha_\lambda U^\lambda|_\beta)|^\beta_\alpha + (b^{\alpha\lambda}|_\beta U_\lambda)|^\beta_\alpha - (b^\alpha_\lambda b^\lambda_\beta W)|^\beta_\alpha]$$

$$= -\nu h^2 \overset{(1)}{S}{}^\lambda_\lambda|^\alpha_\alpha - 2\nu h^3(H\overset{(0)}{S}{}^\lambda_\lambda)|^\alpha_\alpha + [12\Lambda(1+\nu) + 5\nu - \nu_*]\frac{1}{5}h^3(b^\lambda_\mu \overset{(0)}{S}{}^\mu_\lambda)|^\alpha_\alpha$$

$$- \frac{2}{105} \Lambda_* h^5 (b^\lambda_\mu \overset{(0)}{S}{}^\mu_\lambda)|^{\alpha\beta}_{\alpha\beta}$$

$$-3(1+\nu)h[b^\lambda_\mu \overset{(0)}{S}{}^\mu_\lambda + h^2(b^\alpha_\lambda \overset{(0)}{S}{}^\lambda_\beta)|^\beta_\alpha] - 2(1+\nu)h^3(H\overset{(0)}{S}{}^\alpha_\beta)|^\beta_\alpha - (2\nu-\nu_*)h^3(b^\alpha_\beta \overset{(0)}{S}{}^\lambda_\lambda)|^\beta_\alpha$$

$$+\frac{1}{15}\nu_* h^4 \Big[[\overset{(0)}{S}{}^\lambda_\lambda + 2hH\overset{(0)}{S}{}^\lambda_\lambda - hb^\lambda_\mu \overset{(0)}{S}{}^\mu_\lambda]|^{\alpha\beta}_{\alpha\beta} + 2h[b^\lambda_\alpha \overset{(0)}{S}{}^\mu_\mu|^\beta_\lambda + 2b^{\beta\lambda}|_\alpha \overset{(0)}{S}{}^\mu_{\mu|\lambda}]|^\alpha_\beta \Big]$$

$$-\frac{3}{2}(1+\nu)p + \frac{6}{5}[\Lambda(1+\nu) - \frac{1}{2}\nu_*]h^2 p|^\alpha_\alpha - \nu_* h^3[Hp^*|^\alpha_\alpha - (b^\alpha_\beta p^*)|^\beta_\alpha] + \frac{1}{2}\Lambda_* h^3 (b^\alpha_\beta p^*)|^\beta_\alpha$$

$$-\Lambda_* h^4\Big[\Big(\frac{3}{70}p + \frac{1}{15}hHp^*\Big)\Big|^{\alpha\beta}_{\alpha\beta} + \frac{1}{15}h(b^\lambda_\beta p^*|^\alpha_\lambda + 2b^{\alpha\lambda}|_\beta p^*|_\lambda)|^\beta_\alpha\Big] \qquad \textbf{(9.8)}$$

The consistency of relations (9.7) and (9.8) implies a further compatibility condition: if we apply the Laplacian operator to (9.7), and, after multiplying by h^2, we subtract (9.8) from the result, then, we find that the compatibility equation takes the form

$$-Eh^3 \Big[2[(b^\alpha_\lambda U^\lambda|_\alpha)|^\beta_\beta - (b^\alpha_\lambda U^\lambda|_\beta)|^\beta_\alpha] + [(b^{\alpha\lambda}|_\alpha U_\lambda)|^\beta_\beta - (b^{\alpha\lambda}|_\beta U_\lambda)|^\beta_\alpha] - [(b^\alpha_\lambda b^\lambda_\alpha W)|^\beta_\beta - (b^\alpha_\lambda b^\lambda_\beta W)|^\beta_\alpha] \Big]$$

$$= h^2 \overset{(1)}{S}{}^\lambda_\lambda|^\mu_\mu + (1+3\nu-\nu_*)h^3(H\overset{(0)}{S}{}^\lambda_\lambda)|^\alpha_\alpha - (15+10\nu+\nu_*)\frac{1}{5}h^3(b^\lambda_\mu \overset{(1)}{S}{}^\mu_\lambda)|^\alpha_\alpha$$

$$+3(1+\nu)h[b^\lambda_\mu \overset{(0)}{S}{}^\mu_\lambda + h^2(b^\alpha_\lambda \overset{(0)}{S}{}^\lambda_\beta)|^\beta_\alpha] + 2(1+\nu)h^3(H\overset{(0)}{S}{}^\alpha_\beta)|^\beta_\alpha + (2\nu-\nu_*)h^3(b^\alpha_\beta \overset{(0)}{S}{}^\lambda_\lambda)|^\beta_\alpha$$

$$+\frac{2}{15}\nu_* h^5 \Big[[(b^\lambda_\alpha \overset{(0)}{S}{}^\mu_\mu|^\alpha_\lambda)|^\beta_\beta - (b^\lambda_\alpha \overset{(0)}{S}{}^\mu_\mu|^\beta_\lambda)|^\alpha_\beta] + 2[(b^{\alpha\lambda}|_\alpha \overset{(0)}{S}{}^\mu_{\mu|\lambda})|^\beta_\beta - (b^{\beta\lambda}|_\alpha \overset{(0)}{S}{}^\mu_{\mu|\lambda})|^\alpha_\beta] \Big]$$

$$+\frac{3}{2}(1+\nu)p - \frac{3}{5}\nu_* h^2 p|^\alpha_\alpha + h^3[(\Lambda_* + \nu_*)(Hp^*)|^\alpha_\alpha - (\nu_* + \frac{1}{2}\Lambda_*)(b^\alpha_\beta p^*)|^\beta_\alpha]$$

$$-\frac{1}{15}\Lambda_* h^5 \Big[[(b^\lambda_\alpha p^*|^\alpha_\lambda)|^\beta_\beta - (b^\lambda_\alpha p^*|^\beta_\lambda)|^\alpha_\beta] + 2[(b^{\alpha\lambda}|_\alpha p^*|_\lambda)|^\beta_\beta - (b^{\alpha\lambda}|_\beta p^*|_\lambda)|^\beta_\alpha] \Big] \qquad \textbf{(9.9)}$$

Formulae (9.2) and (9.3), when modified by the compatibility equations (9.6) and (9.9), are the constitutive relations for the principal problem complementing the equilibrium equations (4.28): this system is to be solved subject to the edge conditions (5.11).

Rather than derive more unwieldy equations by further substitution, we shall merely indicate the procedure to be followed in the subsequent treatment. We first consider equation (9.9) as a formula expressing the quantity $h^2 \overset{(1)}{S}{}^{\lambda}_{\lambda}|^{\mu}_{\mu}$ in terms of the surface forces, together with other stress and displacement quantities, the latter being "correction" terms arising directly form the curvature. This formula is then used to substitute for $h^2 \overset{(1)}{S}{}^{\lambda}_{\lambda}|^{\mu}_{\mu}$ in the factor with coefficient ν_* in (9.7), thus yielding a derived expression for the invariant $\overset{(1)}{S}{}^{\lambda}_{\lambda}$ in terms of displacement components, surface forces and the "correction" terms consisting of products of the stress and curvature components. When this latter expression for $\overset{(1)}{S}{}^{\lambda}_{\lambda}$ together with the formula for $\overset{(0)}{S}{}^{\lambda}_{\lambda}$ implied by (9.4) are then introduced into (9.2) and (9.3), we obtain the modified form of the constitutive relations for the principal problem.

Further modification is possible if we utilize relations (9.2) and (9.4) to eliminate from (9.3) the correction terms, namely those terms involving products of the $\overset{(0)}{S}{}^{\mu}_{\lambda}$ and the curvature components: in particular, it is desirable to eliminate the correction terms involving derivatives of the $\overset{(0)}{S}{}^{\mu}_{\lambda}$. After neglecting terms of order θ^2, we then have what may be considered the final form of the full constitutive relations, exhibiting the expected feature that the terms involving the higher derivatives of the stresses are associated exclusively with the transverse shear deformability modulus Λ.

10. The Contracted Interior Problem

The rationale for considering a simplified "interior" problem is based on the inference that, when the typical linear dimension on the midsurface is sufficiently large compared with the thickness, then, except in the region close to the edge, the variation of the field quantities, in directions parallel to the midsurface, is quite moderate. This feature, which presumes a comparable moderation in the initial geometrical variation of the midsurface, permits a systematic simplification of the constitutive relations.

It is, therefore, appropriate to introduce a length scale L characterizing the minimum wavelength of variation in both the fundamental form and in the field quantities. For the comparison of this characteristic length with the shell thickness, we introduce a second dimensionless parameter β defined by

$$\beta = h/L \tag{10.1}$$

and the effects associated with the parameter β are comparable with the dominant effects only in the edge zone or boundary layer. Thus the interior problem is characterized by the inequality

$$\beta \ll 1 \tag{10.2}$$

which provides the procedure by which the appropriate simplifications are to be introduced.

In the contraction leading to the equations describing the interior state, we shall omit all effects of order β^2 from the full system of equations governing the principal problem. Since effects of order θ^2 have already been neglected, it will be consistent with the interior approximation to omit all effects of order θ^2, β^2 and $\beta\theta$.

Recalling the left hand sides of relations (9.2) and (9.3), and noting that it is consistent with the above approximation to omit the last term on the left of (9.2), we introduce the measures of deformation $e_{\alpha\beta}$ and $\kappa_{\alpha\beta}$ for the interior problem by setting

$$e_{\alpha\beta} = \frac{1}{2}(U_{\alpha|\beta} + U_{\beta|\alpha}) - b_{\alpha\beta} W \tag{10.3a}$$

$$\kappa_{\alpha\beta} = W|_{\alpha\beta} + b^{\lambda}_{\alpha} U_{\lambda|\beta} + b^{\lambda}_{\beta} U_{\lambda|\alpha} + b^{\lambda}_{\alpha|\beta} U_{\lambda} - b^{\lambda}_{\alpha} b_{\lambda\beta} W \tag{10.3b}$$

where the former measure the mean strains and the latter are termed the mean curvature changes. Imposing the approximations for the interior problem on the constitutive relations (9.2) and (9.3), we see that, with the notation of (10.3), the latter now assume the form

$$\begin{aligned} E\, e_{\alpha\beta} = {} & (1+\nu)\Big[\overset{(0)}{S}_{\alpha\beta} + \frac{2}{3} hH \overset{(1)}{S}_{\alpha\beta} - \frac{2}{3} hb_{\alpha\lambda} \overset{(1)}{S}{}^{\lambda}_{\beta} - \frac{1}{3} hb_{\beta\lambda} \overset{(1)}{S}{}^{\lambda}_{\alpha}\Big] \\ & - g_{\alpha\beta}\Big[\nu\Big(\overset{(0)}{S}{}^{\lambda}_{\lambda} + \frac{2}{3} hH \overset{(1)}{S}{}^{\lambda}_{\lambda}\Big) - \frac{1}{3}(\nu - \nu_*) hb^{\lambda}_{\mu} \overset{(1)}{S}{}^{\mu}_{\lambda}\Big] + \frac{2}{3} \nu h b_{\alpha\beta} \overset{(1)}{S}{}^{\lambda}_{\lambda} \\ & - \frac{1}{2}\nu_*\Big[g_{\alpha\beta}\Big(p^* + \frac{4}{5} hHp\Big) - \frac{4}{5} hb_{\alpha\beta} p\Big] \end{aligned} \tag{10.4}$$

$$\begin{aligned} - Eh\kappa_{\alpha\beta} = {} & (1+\nu)\Big[\overset{(1)}{S}_{\alpha\beta} + 2hH \overset{(0)}{S}_{\alpha\beta} - 2hb_{\alpha\lambda} \overset{(1)}{S}{}^{\lambda}_{\beta} - hb_{\beta\lambda} \overset{(1)}{S}{}^{\lambda}_{\alpha}\Big] \\ & - g_{\alpha\beta}\Big[\nu\Big(\overset{(1)}{S}{}^{\lambda}_{\lambda} + 2hH \overset{(0)}{S}{}^{\lambda}_{\lambda}\Big) - \frac{1}{5}(5\nu - \nu_*) hb^{\lambda}_{\mu} \overset{(0)}{S}{}^{\mu}_{\lambda}\Big] + (2\nu - \nu_*) hb_{\alpha\beta} \overset{(0)}{S}{}^{\lambda}_{\lambda} \\ & - \nu_* g_{\alpha\beta}\Big(\frac{3}{5} p + hHp^*\Big) + \frac{1}{2}(\Lambda_* + 2\nu_*) hb_{\alpha\beta} p^* \end{aligned} \tag{10.5}$$

Bearing in mind that these relations are to be used in conjunction with the equilibrium equations (4.28), we find that further simplification is possible. An inspection shows that if we introduce relations (10.5) for the $\overset{(1)}{S}_{\alpha\beta}$ into (4.28) then the terms thus arising from the zero-th coefficients $\overset{(0)}{S}_{\alpha\beta}$ (and p^*) are of the order β^2, θ^2, or $\beta\theta$ compared with the corresponding terms already appearing in the equilibrium equations. It is, therefore, consistent with the approximation already made to replace (10.5) by the simpler form*

$$-Eh\kappa_{\alpha\beta} = (1+\nu)\overset{(1)}{S}_{\alpha\beta} - g_{\alpha\beta}\left(\nu \overset{(1)}{S}{}^{\lambda}_{\lambda} + \frac{3}{5}\nu_* p\right) \tag{10.6}$$

To invert (10.6) we first take the trace

$$-Eh\kappa^{\lambda}_{\lambda} = (1-\nu)\overset{(1)}{S}{}^{\lambda}_{\lambda} - \frac{6}{5}\nu_* p \tag{10.7}$$

yielding the expression

$$\overset{(1)}{S}{}^{\lambda}_{\lambda} = -\frac{Eh}{1-\nu}\kappa^{\lambda}_{\lambda} + \frac{6}{5}\frac{\nu_*}{1-\nu}p \tag{10.8}$$

which, when substituted into (10.6), given, after a transposition,

$$\overset{(1)}{S}_{\alpha\beta} = -\frac{Eh}{1-\nu^2}\left[(1-\nu)\kappa_{\alpha\beta} + \nu g_{\alpha\beta}\kappa^{\lambda}_{\lambda}\right] + \frac{3}{5}\frac{\nu_*}{1-\nu}g_{\alpha\beta}p \tag{10.9}$$

Returning to the other set of relations, we take the trace of (10.4) and obtain

$$Ee^{\lambda}_{\lambda} = (1-\nu)\overset{(0)}{S}{}^{\lambda}_{\lambda} + \frac{2}{3}(1+\nu)hH\overset{(1)}{S}{}^{\lambda}_{\lambda} - \left(1+\frac{1}{3}\nu+\frac{2}{3}\nu_*\right)hb^{\lambda}_{\mu}\overset{(1)}{S}{}^{\mu}_{\lambda} - \nu_* p^* \tag{10.10}$$

which, on inversion gives for $\overset{(0)}{S}{}^{\lambda}_{\lambda}$

$$\overset{(0)}{S}{}^{\lambda}_{\lambda} = \frac{E}{1-\nu}e^{\lambda}_{\lambda} - \frac{2}{3}\frac{1+\nu}{1-\nu}hH\overset{(1)}{S}{}^{\lambda}_{\lambda} + \frac{3+\nu+2\nu_*}{3(1-\nu)}hb^{\lambda}_{\mu}\overset{(1)}{S}{}^{\mu}_{\lambda} + \frac{\nu_*}{1-\nu}p^* \tag{10.11}$$

Inserting formula (10.11) into (10.4), we transpose the strain terms and, after a rearrangement, we obtain

$$\begin{aligned} &E[(1-\nu)e_{\alpha\beta} + g_{\alpha\beta}e^{\lambda}_{\lambda}] \\ &= (1-\nu^2)\left[\overset{(0)}{S}_{\alpha\beta} + \frac{2}{3}hH\overset{(1)}{S}_{\alpha\beta} - \frac{2}{3}hb_{\alpha\lambda}\overset{(1)}{S}{}^{\lambda}_{\beta} - \frac{1}{3}hb_{\beta\lambda}\overset{(1)}{S}{}^{\lambda}_{\alpha}\right] \\ &\quad + g_{\alpha\beta}\left[\frac{4}{3}\nu^2 hH\overset{(1)}{S}{}^{\lambda}_{\lambda} - \frac{1}{3}(1+\nu)(2\nu+\nu_*)hb^{\lambda}_{\mu}\overset{(1)}{S}{}^{\mu}_{\lambda}\right] + \frac{2}{3}\nu(1-\nu)hb_{\alpha\beta}\overset{(1)}{S}{}^{\lambda}_{\lambda} \\ &\quad - \frac{1}{2}\nu_*\left[g_{\alpha\beta}\left((1+\nu)p^* + \frac{4}{5}(1-\nu)hHp\right) - \frac{4}{5}(1-\nu)hb_{\alpha\beta}p\right] \end{aligned} \tag{10.12}$$

* That (10.5) may be replaced by (10.6) implies further that, within the present approximation procedure, the mean curvature change are insensitive to the addition of strain-curvature terms.

In the right hand side of (10.12), we now employ relations (10.9) and (10.8) to substitute for the elements $\overset{(1)}{S}_{\alpha\beta}$, $\overset{(1)}{S}{}^{\lambda}_{\mu}$ and $\overset{(1)}{S}{}^{\lambda}_{\lambda}$ in terms of the mean curvature changes. We, thereby, obtain the formulae expressing the coefficients $\overset{(0)}{S}_{\alpha\beta}$ explicitly in terms of the deformation measures: a further rearrangement then renders these relations in the form

$$\overset{(0)}{S}_{\alpha\beta} = \frac{E}{1-\nu^2}[(1-\nu)e_{\alpha\beta} + \nu g_{\alpha\beta}e^{\lambda}_{\lambda}]$$

$$+ \frac{Eh^2}{3(1-\nu^2)}\Big[(1-\nu)[2\mathrm{H}\kappa_{\alpha\beta} - 2b_{\alpha\lambda}\kappa^{\lambda}_{\beta} - b_{\beta\lambda}\kappa^{\lambda}_{\alpha}]$$

$$+\nu[(2\mathrm{H}g_{\alpha\beta} - b_{\alpha\beta})\kappa^{\lambda}_{\lambda} - 2g_{\alpha\beta}b^{\lambda}_{\mu}\kappa^{\mu}_{\lambda}] - \nu_* g_{\alpha\beta}[b^{\lambda}_{\mu}\kappa^{\mu}_{\lambda} + \frac{\nu}{1-\nu}2H\kappa^{\lambda}_{\lambda}]\Big]$$

$$+ \frac{1}{2}\frac{\nu_*}{1-\nu^2}\Big[(1+\nu)g_{\alpha\beta}p^* + \frac{2}{5}\frac{1+\nu}{1-\nu}[(1-\nu)b_{\alpha\beta} + \nu 2Hg_{\alpha\beta}]hp\Big]$$

(10.13)

Relations (10.9) and (10.13) are the reduced constitutive relations, for the contracted interior formulation of the first approximation principal problem, complementing the equilibrium equations (4.28).

Neglecting the effect of surface pressure, we now write these relations in terms of the resultants (4.9): we obtain

$$N_{\alpha\beta} = \frac{2Eh}{1-\nu^2}[(1-\nu)e_{\alpha\beta} + \nu g_{\alpha\beta}e^{\lambda}_{\lambda}]$$

$$+ \frac{2Eh^3}{1-\nu^2}\Big[(1-\nu)[2\mathrm{H}\kappa_{\alpha\beta} - 2b_{\alpha\lambda}\kappa^{\lambda}_{\beta} - b_{\beta\lambda}\kappa^{\lambda}_{\alpha}]$$

$$+\nu[(2\mathrm{H}g_{\alpha\beta} - b_{\alpha\beta})\kappa^{\lambda}_{\lambda} - 2g_{\alpha\beta}b^{\lambda}_{\mu}\kappa^{\mu}_{\lambda}] - \nu_* g_{\alpha\beta}[b^{\lambda}_{\mu}\kappa^{\mu}_{\lambda} + \frac{\nu}{1-\nu}2H\kappa^{\lambda}_{\lambda}]\Big] \quad \textbf{(10.14a)}$$

$$M_{\alpha\beta} = -\frac{2Eh^3}{3(1-\nu^2)}[(1-\nu)\kappa_{\alpha\beta} + \nu g_{\alpha\beta}\kappa^{\lambda}_{\lambda}] \quad \textbf{(10.14b)}$$

which with the corresponding form (4.31) of the equilibrium equations, namely

$$N^{\beta\alpha}|_\beta - b^\alpha_\lambda M^{\beta\lambda}|_\beta = 0 \tag{10.15a}$$

$$M^{\alpha}_{\beta}|^{\beta}_{\alpha} + b^{\alpha}_{\beta} N^{\beta}_{\alpha} = 0 \tag{10.15b}$$

are to be solved subject to the contracted Kirchhoff form of the boundary conditions.

Background Survey

The literature on the basic equations of shell theory has grown considerably since the original derivation of the expression for the elastic energy by Aron [1] in 1874, and the associated works of Mathieu [23], Rayleigh [27] and Lamb [20]. With the appearance of the basic and significant paper of Basset [2] in 1890, followed by the more general and comprehensive treatment of Love [21(a)] in 1893, each of which included a discussion of the form taken by the edge conditions necessitated by the Kirchhoff contraction, the foundations of the classical theory were established.

Extending to the shell configuration the Kirchhoff hypothesis that normals to the undeformed midsurface deform, without extension, into normals to the deformed midsurface, Love [21(a)] derived the system of constitutive relations, appropriate for a first approximation description of the principal effects, asymptotically valid in the interior. These relations correspond to, but are not quite identical with, the system (10.14). In emphasizing the significance of the terms involving the curvature changes in the expressions for the stress resultants (eq. (10.14a)), Love [21(a)] acknowledges that the necessity for the retention of such terms had already been recognized by Basset [2], whose derivation, however, had been quite different. In Love's procedure these relations follow from the evaluation of certain integrals, in which, expressions, quadratic in the thickness coordinate, appear in the integrands.

In the rewriting of the second and later editions [21(b), (c), (d)], Love alters the sequence somewhat and first shows that by retaining only the linear terms in the integrands, one obtains a system of uncoupled relations equivalent to (10.14b) together with the relations obtained from neglecting the curvature changes on the right of (10.14a). Having proposed this simpler system as adequate for a wide class of problems, he then proceeds to the derivation of the full expressions for the stress resultants, which would be necessary for a consistent theory covering all cases. The system including these latter expressions then corresponds to the original system of the first edition and Love anticipates that the necessity for the full expressions would arise in problems of inextensional bending.

It is understandable that the attractive simplicity of the uncoupled relations should merit them special attention, particularly since so many problems are insensitive to the inconsistency implicit in their utilization. Apparently because of its association with an integrand linear in the thickness coordinate, Love [21(b), (c), (d)] labels the uncoupled system a first approximation even though he clearly recognizes its inconsistency: the full system, associated with an integrand quadratic in the thickness coordinate, is then in the later editions, [21(b), (c), (d)] described as a second approximation. The nomenclature has proved unfortunate since Love's "first approximation," although admittedly inconsistent, has entered the folklore, while the full system—the only system appearing in the first edition and which, as pointed out by Love, is in substantial agreement with the system derived by Basset—seems to have been largely ignored. Equally unfortunate is the fact that Basset's pioneering work has not been accorded either its due recognition or its deserved attention.

The fourth and final edition of Love's treatise, printed in 1927, includes references to the work of other investigators published in the intevening years. In the following decade there appeared the more restricted investigations of Flugge [8] and Donnell [7]. The suggested refinement of Love's constitutive relations proposed in the former work were later derived in more general form by Lurje [22] and independently by Byrne [4]. Meantime Trefftz [32] adapted the variational method to the shell of general form, while Reissner [29(a)] derived a form of Love's first approximation in a simpler and more elegant manner. The intrinsic theory developed by Synge [31] and Chien [5, 21], while interesting in its treatment of the shell as a two-dimensional continuum, is of marginal interest to the present outline since it is not concerned with relating the shell equations to the three dimensional theory. For a fuller discussion of work prior to 1950, we refer to the report of Hildebrand, Reissner and Thomas [12] and the book by Green and Zerna [11], while the independent developments in the Soviet Union are reviewed by Novoshilov [26].

Further consideration of the constitutive relations based on the Kirchhoff-Love hypothesis were initiated by Reissner [29(b)] and further developed in the joint work with Knowles [17(a), (b)] while the method of asymptotic expansions was applied to the same problem in the paper of Johnson and Reissner [16]. These investigations are further discussed in the survey article of Reissner [29(c)], which also includes an extension of the variational procedure leading to a set of constitutive relations with the further refinement that they include the effects of transverse shear and transverse normal stress. For such systems as the latter, which admit the satisfaction of three independent boundary conditions, the Kirchhoff-Basset contraction is no longer appropriate.

With the renewed recognition of the inadequacy of Love's "first approximation" following the investigation of the helicoidal shell by Cohen [6], there appeared a number of suggested modifications of Love's uncoupled system: in particular we mention the systems suggested by Sanders [3, 30], Koiter [18] and Budiansky [3]. Generally, these involve a judicious modification of the tensor of stress resultants so as to accommodate the symmetric features required of an uncoupled system, while avoiding any obvious violation of the equilibrium conditions. The example of the helicoid [6], which exposes the inconsistency in Love's uncoupled system, is, in fact a case of approximate inextensional bending as anticipated by Love, who warned that in such cases the full system must be used. While the uncoupled systems proposed in [3, 18, 30] circumvent the particular difficulty in the reduction of the specific problem of the symmetrically deformed helicoid, they exclude many terms of the fuller set, thereby admitting possible errors in the calculation of the stress resultants. They also ignore the cautionary remarks of Love and Basset on the necessary retention of such terms in a general approximation scheme designed to cover all possible contingencies.

An interesting comparison of these uncoupled systems together with a discussion of other selected sets of constitutive relations appear in the report of Naghdi [24(b)]. In Naghdi's work [24(b),(c)], there also appears an alternative derivation of the refined set of constitutive relations that includes the effects of transverse shear stress. If, from the relations of Section Nine, one omits the effects of transverse normal stress together with certain other terms, then by introducing Naghdi's assumptions on the displacements, it is possible to derive a set of relations in substantial agreement with his.

Further application of the method of asymptotic expansion is pursued in the papers of Reiss [28], Johnson [15], Green [10], Goldenweizer [9] and Reissner [29(d)]. A summary of the latter paper appears in the survey article of Reissner [29(e)], which also includes a more general discussion of the possible approaches to the problems of shell theory. In the context of expansion procedures, we also note the work of Hu [13], whose procedure, though substantially different from that followed here, is also based on expansions in terms of Legendre polynomials.

The interesting work of John [14], whose main thrust is at the nonlinear theory, shows how the technique of Sobolev can be applied to derive estimates on the relative significance of the quantities in the three dimensional theory, whereby one can extract the dominant effects: the procedure leads to a pair of coupled two-dimensional equations with estimates on the neglected terms. While this significant investigation may ultimately be the basis for establishing a rigorous foundation for the two-dimensional theory, it is incomplete in its present form: the procedure manages to avoid the issue of the constitutive relations and leaves some ambiguity on the question of a first approximation. Also in 1965 there appeared the comprehensive analysis of Vekua [33] aimed at the systematic derivation of the successive approximations of shell theory. There, too, the procedure is based on expansions in terms of Legendre polynomials. However, for his fundamental elements Vekua takes the displacement components, whose Legendre representations are then introduced into the three-dimensional equations of elasticity. This work includes a treatment of such factors as thickness variation and discusses some of the general analytic procedures developed by the author.

For an indication of some later trends and developments, we refer to the Proceedings of the 1967 IUTAM Symposium on Shell theory edited by Niordson [25(a)], to the Handbuch der Physik exposition by Naghdi [24(d)], to the volume by Koiter and Simmonds [19], and also to the more recent work by Niordson [25(b)].

Apart from the basic difficulties of the problem, it appears that more confusion than simplification has resulted from not maintaining the distinction between the curvature parameter (θ^2) associated with the shape and the less clearly defined parameter β^2 associated with the edge-effects. It should be noted that while there is every reason to expect convergence for the approximations with respect to θ^2, any expansion in powers of β^2 can, at best, be asymptotic. Moreover, many investigations have been based on the premise that $\theta\,(=h/R)$ is the appropriate approximation parameter – a tradition that seems to claim its origin in the work of Love [21], though it does not appear to be explicitly stated there. In our method of integration, the quantity θ^2 has emerged as the appropriate thickness-curvature parameter, and any effort at considering θ as the basic approximation parameter would have to be tied to an extra assumption on the relative magnitude of the bending and stretching effects that would compromise the inclusion of a state of inextensional bending.

These features are further confounded by the incidence that there is a significant class of problems for which, in fact, the relation $L = \sqrt{hR}$ holds so that the parameters β^2 and θ become indistinguishable. The success of a procedure, based on neglecting effects of order θ, in those cases, is due to the fact that, in deriving the leading term in an asymptotic expansion, it is consistent to neglect effects of order β^2. Apart from the fact that such coincidence cannot be expected in general, the distinct nature of the two parameters requires that they be treated separately.

In view of the heavy volume of published material, the above survey must necessarily remain but a selective outline of the vast literature, citation to which may be found in the listed references.

References

1. Aron, H, J.f. Math., (Crelle), Bd. 78, 1874.
2. Bassett, A. B., Phil. Trans. Roy. Soc. (Ser. A) vol. 181, 1890.
3. Budianski, B. and Sanders, J. L., Prager Anniversary Volume, 1963.
4. Byrne, R., Univ. Calif. Publ. Math., N.S. 2, 1944.
5. Chien, W. Z., Quart. Appl. Math., vols. 1 and 2, 1944.
6. Cohen, J. W., Proc. IUTAM Symp. on Shell Theory 1959, (W. T. Koiter, ed.), Amsterdam, 1960.
7. Donnel, L. H., (a) N.A.C.A., T.R. No. 470, Washington, D.C., 1933
   (b) Proc. Fifth Intern. Cong. Appl. Mech., New York, 1938.
8. Flugge, W., Eng. Arch., vol. 3, 1932.
9. Goldenweizer, A. L., (a) Theory of Elastic Thin Shells, (Moscow 1953), New York, 1961.
   (b) Appl. Math. & Mech. (PMM), vols. 26 and 27, 1962-63.
10. Green, A. E., Proc. Roy. Soc., Ser. A, vol. 266, 1962.
11. Green, A. E. and Zerna, W., Theoretical Elasticity, Oxford, 1954.
12. Hildebrand, F. B., Reissner, E., and Thomas, G. B., N.A.C.A., TN, 1833, Washington, D.C., 1949.
13. Hu, W. C. L., Tech. Rep. No. 5, NASA Contract, NAS – 94(06), Sw. R.I., San Antonio, Texas, 1965.
14. John, F., Comm. Pure Appl. Math., vol. 18, 1965.
15. Johnson, M. W., J. Math and Phys., vol. 42, 1963.
16. Johnson, M. W., and Reissner, E., J. Math. and Phys., vol. 37, 1958.
17. Knowles, J. K. and Reissner, E.: (a) J. Math and Phys., vol. 35, 1956-57.
   (b) J. Math. and Phys., vol. 37, 1958.
18. Koiter, W.T., Proc. IUTAM Symp. on Shell Theory 1959 (W. T. Koiter, ed.), Amsterdam, 1960.
19. Koiter, W.T. and Simmonds, J.G.: Foundations of Shell Theory, Proc. 13th Int. Cong. Theor. and Appl. Mechs., Springer-Verlag, 1972.
20. Lamb, H., Proc. London Math. Soc., vol. 21, 1891.

21. Love, A. E. H., (a) A Treatise on the Mathematical Theory of Elasticity (2 vols.), Cambridge, vol. 2, 1st Ed., 1893.
(b) ibid., (single volume), 2nd Ed., 1906.
(c) ibid., (single volume), 3rd Ed., 1920.
(d) ibid., (single volume), 4th Ed., 1927.
22. Lurje, A. I., (a) Prikl. Mat. Mek, vol. 4, 1940.
(b) Appl. Math. & Mech. (PMM), vol. 12, 1950.
23. Mathieu, E., J. de l'Ecole Polytechnique, t. 51, 1883.
24. Naghdi, P. M., (a) Quart. Appl. Math., vol. 14, 1956.
(b) Tech. Rep. No. 15, Inst. of Eng. Res., Univ. of Calif., Berkeley, 1962.
(c) Progress in Solid Mechanics, 4, New York, 1963.
(d) Theory of Plates and Shells, Handbuch der Physik, vol. 6A/2/, Springer, 1972.
25. Niordson, F. I., (a) Proc. IUTAM Symp. on Shell Theory 1967, Copenhagen, 1969. (Editor)
(b) Shell Theory, North Holland, Amsterdam, 1985.
26. Novoshilov, V. V., The Theory of Thin Shells, Leningrad, 1951.
27. Rayleigh, Lord: Proc. London Math. Soc., vol. 13, 1882.
28. Reiss, E. L., Quart. J. Mech. Appl. Math., vol. 15, 1962.
29. Reissner, E., (a) Amer. J. Math., vol. 63, 1941.
(b) J. Math. and Phys., vol. 31, 1952.
(c) Proc. 1st Symp. Naval Structural Mech. (1958), 1960.
(d) J. Math. and Phys., vol. 42, 1963.
(e) Proc. Eleventh Intern. Cong. Appl. Mech. (Munich, 1964), Berlin, 1965.
30. Sanders, J. L., NASA Report R-24, Washington, D.C., 1959.
31. Synge, J. L., and Chien, W. Z., Von Karman Anniv. Vol., 1941.
32. Trefftz, E., Z. f. angew. Math. Mech., vol. 15, 1935.
33. Vekua, I. N. (BEKYA, Й. H.), Theory of Thin Shells of Varying Thickness, (in Russian), Tbilisi, 1965.